THE ADVENTURER'S SON

The

ADVENTURER'S SON

A Memoir

ROMAN DIAL

WILLIAM MORROW
An Imprint of HarperCollins*Publishers*

HarperCollins books may be purchased for educational, business, or sales promotional use. For information, please email the Special Markets Department at SPsales@harpercollins.com.

FIRST EDITION

Designed by Bonni Leon-Berman

Library of Congress Cataloging-in-Publication Data has been applied for.

ISBN 978-0-06-287660-7

20 21 22 23 24   LSC   10 9 8 7 6 5 4 3 2 1

Trial and error,

Failure and terror,

The truth of the matter at hand.

Death in a whisper

Is so much to weather

For the life of a wife

And her man.

CONTENTS

Prologue: Family

Tide-pooling on the northern California coast, 1989.

MY WIFE, PEGGY, bore our son, Cody Roman Dial, on February 22, 1987, in Fairbanks, Alaska. Peggy and I had met as teenagers there, a place that drew me for its prospect of climbing mountains, skiing glaciers, and rafting rivers. We purposefully raised Cody Roman and his sister, Jazz, in Alaska, exposing them both to natural history travel and wilderness experiences around the world. When he was six, Cody Roman and I hiked alone across sixty miles of an empty Aleutian island. When he was nine, we made a family trip to a remote national park in Indonesian Borneo, a life-changing experience in an amazing tropical rainforest. While activities like wilderness trips and travel in the developing world set him apart from many kids, Cody Roman played with Legos and video games, listened to indie rock, read Harry Potter, and attended public school as a normal kid from a close-knit family brought closer by nature and adventure. As a professional scientist and explorer, I included him in science and explorations as my favored and willing partner in both.

At twenty-six, Cody Roman left graduate school in Alaska for three months on the East Coast, then six more in Latin America. He explored volcanoes, rivers, ruins, reefs, and jungles on his own and with other travelers he met along the way. As he traveled, he stayed in touch with friends and family, emailing plans, maps, and stories. Then in July 2014, while he was in Costa Rica—after sending details of a planned five-day walk, alone and off-trail—the emails ended abruptly. Alarmed and guilt-stricken, I fought down rising panic and rushed down to find his trail before it was too late.

This book is the story of our lives and my search for my son. Some of the dialogue was unforgettable; some transcribed; some told as oral history for decades; some imagined. It has been a painful book to write: full of nostalgia, catharsis, sadness, longing, and struggles with guilt. But the story is an important one—the most important words I have ever written. I owe it to Cody Roman to get it right and true.

PART I

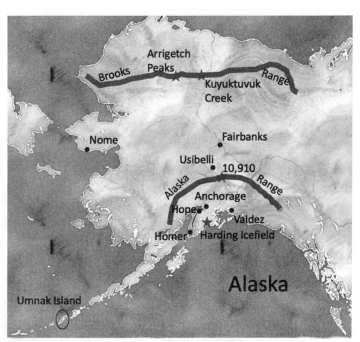

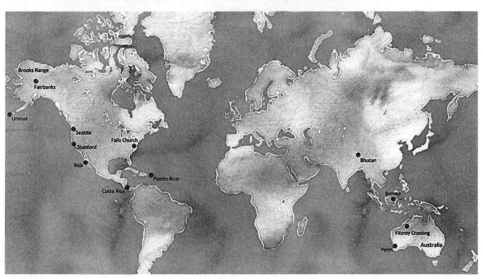

CHAPTER 1

Usibelli

Author and author's uncle, Brian Decker. Rochester, Washington, 1973.

In 1955, a sixteen-year-old girl named Linda Eklund left her family's ten-acre farm near Rochester, Washington, to live in Seattle, probably to escape her stepfather, or maybe her mother's gruff Germanic way. Linda met my father at twenty, fell in love, and gave birth to me at twenty-one. Four years later she had my sister, Tamara.

"Was I a mistake?" I remember Tamara once asking.

"No. Your brother was, though," my mom quipped back. I thought through what it meant to be a mistake, feeling a bit of a sting. Sensing my disappointment, my mom went on: "But your dad liked him so much he wanted another one, and that was you, Tamara Dial." My mom named Tamara after her best friend, who'd helped her get on her feet once she'd left home.

My father named me after his uncles—Roman and Joseph, born in Poland—who'd been father figures to him on their farm in Enumclaw, east of Seattle. My dad never met his own father, nor, as a self-confessed city slicker, did he take much to the outdoors. After meeting my namesakes, who were somewhat distant and unaffectionate, I came to understand why my dad struggled as a father to me: he wasn't sure how to be one.

Like all boys, I was fascinated by my father and drawn to him like a moth to a bare bulb in the dark, watching and studying and learning what I could. The fondest memory I have of Bob Dial came from February 1970 when I was nine. My dad, a Ph.D. civil engineer who developed computer models to describe traffic flow, had taken a job in northern Virginia. While Tamara and my mom flew east to our new home in Falls Church, he drove our Shetland sheepdog, Brute, and me across the country in our Porsche Speedster.

It was a marvelous trip, twisting down the Oregon coastline, under the California redwoods, over the Sierras and Rockies, then across the empty plains of Kansas and the flat hardwood forests east of the Mississippi. We talked and watched the continent roll by and sometimes he sat me on his lap to steer the silver Speedster down curvy country roads. That trip holds warm memories of my dad and we bonded then. Later I would learn that bonds need maintenance to endure.

In May 1970, my parents bought me a ticket to Alaska, where I would stay with my mother's brothers in Usibelli, a mining camp in the Alaska Range. At the time, the trip seemed a substitute for my friends' adventures at summer camp. As an adult, though, it occurred to me that my parents sent me away while they struggled with their marriage. In my mom's drawer of old photos, there are none of Bob Dial with our family after that summer. Tamara and I would see him only on weekends, when he was often late to pick us up. Sitting in our house, we waited, disappointed he was more interested in his life than in ours.

The summer I headed to Usibelli, my parents' problems were invisible to me, a little kid who knew only that Alaska would be even more exciting than his grandmother's farm. My grandmother lived an hour and a half from Seattle, with a dozen head of cattle, pigs, rabbits, vegetable gardens, and blackberry brambles. Exploring the surrounding countryside and discovering its wildlife made the zoo in Seattle feel like mere spectating. My uncles—Zinn and Brian—were kind to me, their big sister's son, a skinny, precocious city kid with no common sense, as both were quick to remind me with a laugh. They taught me lessons about nature and life that no classroom or books could offer.

After picking me up at the Fairbanks airport, Zinn drove south with me in the back seat, nose to the window, soaking in the view. It was my first trip to Alaska and already I was intoxicated by the midnight sun and the landscapes uninterrupted by buildings, fences, or anything man-made beyond the gravel road. Three hours later, he turned his Ford pickup off the Parks Highway and headed east toward Healy.

Zinn drove slowly to keep the dust down as we passed woodlands of stunted spruce and dwarfed aspen covering the foothills of the Alaska Range. He guided the Ford across the one-lane trestle of a railroad spur leading to the coal mining district at Usibelli. I looked under the railroad ties at the bossy Nenana River, its glacial gray waters writhing, hypnotic and terrifying.

Beyond the bridge the road twisted past soft cliffs smoking with burning coal seams. To the south, scabby peaks rose above pale tundra, their summits cradling winter's lingering snow.

My uncles lived and worked at the Usibelli coal mine. A collection of scattered sheet metal and wood clapboard buildings set among tractor-truck trailers, Usibelli itself was barely a company town for the Usibelli Coal Company. Both uncles worked long hours operating heavy machinery that stripped coal from rolling hills. While my mom had sent me there under their care, it was clear Brian and Zinn were busy. It would be up to me to entertain myself. Luckily there was plenty to do under the benign neglect of my uncles.

Exactly nine years older, Brian shared my birthday. Kid-sized and kid-hearted, with bright blue eyes beneath brows that seemed always arched in amusement, Brian sometimes stuttered, but his staccato statements merely emphasized what he tried to spit out. Maybe because he was the baby in his family and I was younger—but old enough to be a brother—he introduced me proudly to his friends as "my little nephew." Like Zinn, he often called me "Rome."

"Hey, Rome!" Brian grinned as Zinn carried my bags into Brian's bunkhouse room my first night in Usibelli. "You can sleep here. Zinn and me gotta work tomorrow but we'll try and take you out on Zinn's Kawasaki this weekend." Zinn, who had brought his wife, Faye, their three-year-old son, and infant daughter to Usibelli, stayed in a house next door. Faye was supposed to keep an eye on me, but she rarely did.

Brian gave me a quick lesson on how to survive in the bunkhouse, empty all day while everyone was off stripping coal. "This here's the oven. And here"—he opened the freezer—"are the Tater Tots. Just turn on the oven, put the Tots on the cookie sheet, and cook 'em up until you can smell 'em. Eat whatever you want, but don't b-b-burn down the bunkhouse!" he instructed with a laugh. "Now, if you're gonna leave camp," he said, turning serious, "take Moose with you. See ya tonight, Rome!" And with that he left for work and I left with Moose to explore.

Moose was the camp dog. Zinn claimed Moose was half wolf, and I believed him. His coat was thick, unlike that of any dog I had ever petted, and he was tall, with long, lanky legs and big feet on an otherwise German shepherd frame. He wagged his tail and looked at me with a dog smile when I rubbed his back.

There were no computers or televisions in rural Alaska in 1970. In their place, I had books and a taxidermy correspondence course, a .22 rifle my uncles entrusted me with, and a Kawasaki dirt bike that was too big for me. The motorcycle's front brake lever was broken in half, the result of a kick-start failure. To kick-start the bike required that I launch my skinny frame up with both feet off the ground, shove the kicker down with my right foot to fire the ignition, then engage the clutch and first gear, all before the bike fell over. It didn't always get going in time. When it did, I toured the mining roads, thrilled to drive on my own; but I grew bored just zooming around.

Most of my favorite explorations were on foot and off-trail with Moose out front. We pushed through willows and alders, rock-hopped, waded streams, and explored two nearby ghost towns called Suntrana and Lignite, where the coal had been mined out but the scent of diesel still lingered. Wood frogs waited in tundra ponds, magpies in shrub thickets, red squirrels in boreal woods. I carried Peterson's field guides to identify the birds and mammals. The books nourished my dream of growing up to be a scientist; the nature near Usibelli gave my dream form.

In early autumn, Zinn took me on a bowhunt for moose off the Stampede Trail. The midnight sun was gone and it got dark at night with the northern lights shimmering overhead. We left early to find a moose, and my toes didn't thaw until the frost had melted off the red leaves of fireweed. I tried to be as quiet as I could, but Zinn looked back at me. "You sure are noisy, aren't you?" His big fake teeth flashed in a smile. His real ones had been knocked out during a fight with his best friend.

I doubled down on not stepping on any sticks, not brushing noisily against any bushes, and certainly not talking. I kept close behind, carrying Zinn's eight-pound .30-06 rifle that he said we might need for bears. Zinn spotted a brown bulk that we stalked quietly together until he asked me to wait while he went ahead. I sat patiently cradling the rifle. Careful with its scope as Zinn instructed, I watched bugs crawl and leaves fall.

Then Zinn appeared mysteriously out of the brush. "It's a cow," he whispered, aware there might still be a bull nearby. We could only take a male moose, so we continued our hunt until the pungent odor of high-bush cranberries hung thick in the woods. "Moose lay down now 'cause it's too hot for 'em. We won't have any luck. Let's go back to Usibelli." Zinn's lessons were based on his humble farm roots and a trip sailing around the world in the U.S. Navy.

Later that fall, near remote Cody Pass, Zinn got a bull caribou without me. He brought back its sweet-tasting meat and its antlers still covered in the short fur called velvet. Like skin, velvet has blood vessels that nourish and grow the antlers until they reach their full size. Then the bull scrapes the velvet off to display his white rack to cows and other bulls, or sometimes to fight for the right to mate.

Zinn asked me to mount the antlers. My taxidermy course materials instructed that flushing them with gasoline would expel the velvet's blood. I doused the furry horns in gas, worked dry preservative between the skull plate and skin, then fastened the rack to a plaque. For my flight home, we boxed the caribou antlers up with a big, black raven I had mounted: two priceless souvenirs from a summer of doing whatever I wanted, learning independence and responsibility.

THE FOLLOWING SUMMER in Virginia, a friend's dad took us to the Appalachian Mountains. We were going to climb Old Rag, a barren granite dome in Shenandoah. "There they are," the father announced when the hills came into view. "The Blue Ridge Mountains!"

"Those aren't mountains!" I said, my appreciation for scenery forever spoiled by Alaska. "Those are just foothills. There's not even any snow on 'em!"

Pronouncements like that—and telling the kids at school about my summer in Alaska with a .22, a wolf-dog named Moose, and a motorcycle—didn't gain me many friends. "Stop bragging, Roman!" they'd say. But their criticism never curbed my enthusiasm for Alaska, with freedoms and adventures impossible back east. My Alaskan experiences gave me the confidence to try anything and the strength to endure my parents' breakup, which had begun with their separation in 1970 and ended in divorce four years later when I was thirteen.

After my parents' marriage ended, my mom married a gentle lawyer from Virginia named Lew Griffith. Although we never called him "Dad," Lew was an excellent father figure to my sister and me. My mom and he nourished my preteen fascinations with milk snakes and plethodontid salamanders, steaming geysers, and sphagnum bogs. She even encouraged my suggestions for family vacations, where I chose the destinations and planned the trips.

Informed by AAA maps and *National Geographic* articles, I charted far-flung natural history tours. With my mom or Lew behind the wheel of the family station wagon, we made road trips in search of colorful amphibians in the Appalachians and insect-eating plants in southern swamps. We cruised summer blacktop at night across Arizona deserts looking for reptiles. My mother even drove Tamara and me across the country to my grandmother's farm on a tour of national parks.

Tamara usually stayed back at the motel pool or with my mom and Lew while I went to look for creatures alone or with Mike Cooper, my best friend, who often joined us on these trips. More interested in dogs and horses, Tamara shied away from the mud, bugs, and spiderwebs that we budding scientists were willing to endure.

Back in the sixties and seventies a boy could still run off to play alone during the idyllic era between rural-agrarian America, when kids worked the land, and today's suburban-urban America, when kids embrace indoor entertainment. The suburbs then, like Holmes Run Acres where we lived in Falls Church, often dove-tailed with natural ecosystems. The Chiles Tract, the last large parcel of undeveloped land inside Washington, D.C.'s Beltway, was only two blocks from my home. I'd spend hours wandering its forests, creeks, and swamps, learning my way in the woods.

Mike Cooper and I filled steamy terraria in our rooms with pink lady-slipper orchids and bright green sphagnum mosses we found in the Chiles Tract. Our bubbling aquaria housed red-spotted newts from the swamps and spotted turtles from the creek. After an escaped snake found its way into my mom's underwear drawer, she reminded me politely but frequently to keep my bedroom door shut.

Our mom valued education and sent Tamara and me to a small progressive private elementary school near our home, where sensitive science and English teachers leveraged my fascination with science and nature into essays and research projects for their classes. But as puberty advanced, my interests shifted away from the nurture of reptiles to the nature of girls.

By my senior year in public high school, my participation in the adventure sports had eclipsed my studies of natural history. With its thrilling physical problem solving far above the ground, rock climbing engaged me most and I took up with two teens who climbed at a high standard: Dieter Klose and Savvy Sanders. After graduation, we three headed to Colorado in Dieter's white Econoline van. I went farther, exploring the West by thumb and freight train, catching the ferry to Alaska at summer's end.

With my good grades and strong interest in adventure sports, my parents and neighbors encouraged me to apply to Princeton and Dartmouth. But I couldn't. Three summer trips to Alaska had—as Henry Gannett of Harriman's 1899 Alaska Expedition warned—dulled my "capacity for enjoyment by seeing the finest first." There was only one place for me.

To my parents' likely dismay, although they never tried to dissuade me, I applied to a distant college with the Alaska Range's sheer Mount Huntington on its catalog cover—UAF, the University of Alaska Fairbanks—to study science, pursue adventure, and once again do whatever I wanted.

Heading north at sixteen, I was too naïve to appreciate the prestige that an Ivy League education would bring. But even now, writing this as an old man near sixty, none of my regrets center on moving to Alaska then.

CHAPTER 2

10,910

10,910, east face.

By the time I arrived on UAF's campus as a college freshman in fall 1977, I considered myself a climber, giving up money, relationships, and social prestige for the thrill of steep terrain. And although not yet an "alpinist," a mountaineer who chooses only direct routes up pointy peaks, I wanted to be one. Starting in high school, that desire would form the backbone of my Alaskan dream for nearly a decade.

Seeking out other climbers, I soon found their cultural center at the Sandvik House, the middle apartment of a three-plex, four blocks from UAF. Climbing the stairs to an open door, I walked into a full-tilt Sandvik party. The Allman Brothers blared on the stereo. Marijuana smoke filled the air. Tapestries hung on the living room walls. Students, vets, and locals sprawled across beanbags or stood huddled together, drinking beer, smoking dope, telling stories, lies, secrets, and jokes. Topographic maps of the central Alaska Range—the local Hayes Range—plastered one side of the hallway leading to the bedrooms. A row of pot plants in five-gallon buckets lined the other. Grow lights made map reading easy. The pot made it fun.

The partygoers were a who's who of Fairbanks climbing. In the living room, a Harpo Marx look-alike held court. Carl Tobin, then twenty-three, with a sharp wit and a boldness unmatched in Alaska, was both popular and central to Fairbanks's new alpinism. Three months later, he would make the first ascent of frozen Bridal Veil Falls, a 600-foot waterfall outside Valdez that was more sustained than any ice climb Alaskans had previously attempted.

By February 1978, another UAF student and I had practiced enough on local waterfalls that we attempted Bridal Veil ourselves. Well after dark, we finished the longest, hardest climb either of us had ever done. A week later, following a slideshow on campus, Tobin hailed me: "Hey, I heard you climbed Bridal Veil." He grinned and I nodded back, starstruck but pleased that he knew who I was and what we had done.

Less than a month later, a group of experienced ski-mountaineers invited me on a ten-day trip into the Alaska Range's Great Gorge of the Ruth Glacier. Flying out of Talkeetna, a skiplane delivered us to the base of Denali. We climbed a minor peak, then skied sixty miles to the highway over sparkling glaciers beneath towering walls. Even though I was by far the worst skier in the group, that ski tour remains one of the few experiences of my life that surpassed all my expectations: simultaneously far more intimidating and alluring than I could have ever imagined. It was my first mountain wilderness expedition. I wanted more challenges like that to both dwarf and empower me.

The technical skills from climbing ice combined with the survival skills of mountain travel provided me with the ingredients to climb alpinist lines and the confidence to lead others on Alaskan adventures. During the summer between my freshman and sophomore years, my high school friend Savvy Sanders joined me for three weeks in the Great Gorge. The year after that, I organized a month-long expedition to the Arrigetch Peaks in the Brooks Range, a trip that taught me painful lessons in group dynamics.

Viewed in photos from books and magazines, the Arrigetch mesmerized me with their geometric summits in a wilderness setting. I spent the winter planning our trip and imagining the routes, especially the most dramatic spire, Shot Tower. I invited Savvy, who convinced Dieter to go. In turn, Dieter enlisted his climbing partner Mike Bearzi, at twenty-five an adult by comparison; both my friends were twenty and I eighteen. After school was out, I hitchhiked to Yosemite to meet Dieter and climb with him for a few weeks.

Dieter and I never got along well, not in Virginia, not in Yosemite, and certainly not in the Arrigetch. Knee-deep in a Yosemite Valley stream, he was nearly washed off the edge of a 600-foot waterfall before I pulled him to safety. Dieter claimed that I had saved his life but he seemed ungrateful for that help while in the Arrigetch, where our Nietzschean power struggle felt straight from *Lord of the Flies*.

During our August in the Arrigetch, it rained most days. The unclimbable weather kept us tent-bound, where we read books, smoked dope, played vindictive games of hearts, and made sure the Cadbury bars were evenly split. Stronger, smarter, funnier, a better climber, and there with his close friends, Dieter soon made all the decisions and criticized any of mine. My dream expedition had morphed into little more than a glorified camping trip with someone who wouldn't speak to me without a sharp word. I had learned that if a relationship is weak to begin with—like mine had been with Dieter—then it will only get worse when stressed. Most important, companions matter more than goals and objectives.

Besides my desire to be an alpinist, I saw myself as a field scientist one day and declared wildlife management my major at UAF. My dad had encouraged me to take a math class every semester in college. He knew that mathematics, the lingua franca of science, would allow me to change subjects later on. Wildlife's curriculum had no room for math but it did require ecology, the science of organisms and their environment. An ecology course at UAF put a name to my lifelong interest in nature and defined my future career.

The science of ecology had been mathematicized in the seventies by the late Princeton University ecologist Robert MacArthur. MacArthur's ecology was informed by his M.S. in mathematics and Ph.D. in biology. Finding ecology allowed me to flourish academically. It offered a science for quantitative naturalists like me, so I switched majors to biology and took every ecology class that UAF offered, adding mathematics as a second major.

Working toward degrees in biology and math left me "a mathematician who wanted to be a biologist" in the view of the biology department. In contrast, faculty in math gave me access to paper-grading jobs that honed useful quantitative skills. With each succeeding math class, the technical papers published in *The American Naturalist* and *Theoretical Population Biology* grew ever more accessible to me. And with each succeeding mountain adventure, Alaska's wilderness did, too.

IN FEBRUARY 1980, I was flummoxed to find Carl Tobin grinning at my cabin door. He came in, kicked off his boots, pulled a couple of beers from his pack, and showed me a photo of a slender white peak atop a sweeping blue-ice wall. It looked like Mount Huntington's cute little sister.

"What a peak," I gushed. "What is it?"

"The east face of Ten Nine Ten," Carl said. A 10,910-foot-high tetrahedron, the mountain is known to Fairbanks climbers simply by its digits. "What do you think? Want to climb it?"

Although I'd made a few glacier ski trips, rock-climbed in the Arrigetch, and ice-climbed in Valdez, I was still just a pimply-faced teen, bruised from what felt like a failed expedition the year before. "Well . . . um." I gulped. "How?"

"Right here." His finger traced directly up the center of the icy blue face, through a rock band, to the airy summit—a true alpinist's line.

The best climber in Fairbanks was asking me to climb the kind of route I'd dreamed of climbing since I was fifteen. But like a shy high school nerd invited to the Sadie Hawkins dance by the captain of the cheerleading squad, I couldn't say yes.

"Pretty sweet, huh?" he coaxed.

"Oh, man. I'd love to," I said, "but," remembering Dieter, "I don't think I'm ready for that."

"You climbed Shot Tower. And the ice won't be any harder than Bridal Veil."

Grateful for his confidence, but aware of my job and school, I asked, "How long would it take?"

"The climb? If conditions are good, a day up and a day down. The whole trip? Two weeks. Fly in and ski out." He grinned. "Think about it."

Within a week I had tracked Carl down to tell him yes. Less easy was approaching my boss, who had been an Alaskan climber, too. He once told me, "All my partners either died or quit climbing." Asked if I could go to the Hayes Range for two weeks, he responded, "Yeah, you can go, but you won't have a job when you get back."

I made my choice and sharpened my tools, shopped for food, and trained at the gym over the weeks that followed. From the gymnasium balcony, a pretty, blond girl watched Carl, me, and other gym rats climb hand-over-hand up twenty-five-foot ropes dangling from the ceiling. She looked young and petite, with a broad smile and high cheekbones below almond-shaped eyes. *She must be here to see Tobin with his shirt off,* I thought.

A week later, Carl and I flew into the Hayes Range. My share of the charter cost my last paycheck. The pilot dropped us off directly below Ten Nine Ten. Its 3,000-foot face looked short, easy, almost disappointing in the trick of perspective known as foreshortening. Years later Carl would tell me, "If it wasn't for foreshortening, nothing would ever get climbed."

In the morning, surrounded by steep, glacier-draped mountains, we headed up. Midway on the route, Carl led through near-vertical granite mixed with ice. He pounded in a warthog—an ice piton that looks like Macbeth's dagger—for protection. Following him, my crampons screeched like fingernails on a chalkboard. I stopped to wrench the warthog out of the crack for use higher up. "Leave it in!" Carl called down, "We gotta keep moving!" With cold toes and only halfway up the climb, I happily left the 'hog skewered in an ice-choked crack, unclipped the carabiner and sling, and hurried on. Pitch after pitch of ice led to the final headwall as the weather slipped into storm.

The increasing snowfall stopped us on a small arête where we would dig in for the night. Carl shoveled a shallow snow cave that sheltered us above the waist only. We crawled into our bivy sacks, sleeping fitfully and cramped, our feet dangling over the edge. Spindrift powdered our faces till dawn. Although I was miserable, it was precisely the experience I'd hoped for—Alaskan alpinism.

The storm passed, leaving the morning clear and cold. On the windless summit, we took in the view while Carl melted snow on our stove for hot cocoa. After my fear of failure, success tasted especially sweet, but we still needed to get down. We descended a sharp ridge back to camp, then packed up and skied out a series of glaciers to the road, elated with our ascent of the virgin face. Carl had won a local prize in Ten Nine Ten, picking a plum with an inexperienced, teenage kid.

I felt special to climb the route with Carl. But one night, years later at a Sandvik party, he admitted, "I would have climbed Ten Nine Ten with anyone." Seeing my face fall, he added quickly, "But I'm glad I did it with you." Carl valued his partners' feelings as much as he valued their belays and we would make many adventures together. Intense outdoor experiences either strengthen or extinguish bonds between partners.

Back in Fairbanks after our climb of Ten Nine Ten, we quickly sobered up from the alpine high. Two friends, a popular couple, had been involved in a tragic fall down a local mountain. Peter McKeith, a UAF graduate student, was the president of the Alaska Alpine Club. His girlfriend was the strongest female climber in Fairbanks. With three broken limbs and trapped between crevasses, she overnighted in her backpack's built-in bivy sack and survived. Peter didn't.

Every adventure community feels elation and sorrow. But for me, springtime's sunshine amplified my elation, burned off much of the sorrow, and prompted me to approach for the first time the pretty eighteen-year old freshman who'd watched us climb ropes in the gym: Peggy Mayne.

CHAPTER 3

Peggy Mayne

Peggy, Brooks Range, July 1986.

Bathed in sunlight the day after classes ended, Peggy Mayne stood at the top of the UAF library stairs. It was warm—a beautiful spring day. We had never met, but we both had anticipated this moment. Peggy, a friend named Eleanor had recently informed me, had been eyeing *me* all winter. "But," Eleanor went on, "she's not your type, Roman."

Seeing Peggy standing there, I thought, *Let me be the judge of that.*

Success on Ten Nine Ten had emboldened me to speak to

her, although forty years later we still disagree about who said hi first. Slim in white painter overalls, she wore her long blond hair straight and loose down her back. Her smile sparkled bright as the May sunshine. When her blue eyes spotted in green met mine, we felt a mutual attraction that was immediate, physical, and uninhibited.

I asked her out on a date to a campus play that night. While I was sitting next to her, any little touch of our elbows or knees so electrified me that I could hardly follow the plot. Afterward, we walked and talked well past midnight, when the sun dips briefly only to rise shortly after. While Peggy Mayne wasn't interested in climbing, she certainly seemed interested in me.

The week following semester's end, we spent every day together. We walked through the woods behind the university. We cycled Fairbanks's dusty bike paths. And we hung out at her sister and brother-in-law's place three blocks from campus. In Maureen and Steve's small house Peggy cut my hair. The intimacy thrilled me as she pushed her taut little body against mine to trim my shoulder-length mane to the collar.

Like me, Peggy liked to talk. We talked while we walked. We talked while we bicycled side by side (neither of us had a car). And as weeks turned to months and we found ourselves in bed, we talked there, too.

Peggy's nature rubbed off on my dirt-bag climber cheapness. She taught me to share, to consider others, to be responsible for myself. Then, as now, she lives her every breath by the Golden Rule. She also shows no mercy when she's unhappy with my behavior.

As the youngest of ten children—five boys and five girls—who grew up under an abusive father, she has always shown remarkable insight and empathy. It's as if her whole nervous system focuses outward, collecting data on others, observing and responding to them. She sees things I don't even know exist in people, judging them on character, not accomplishments.

That summer, I left for Colorado to climb, then for Virginia to work. Peggy left for a salmon cannery on the Alaska Peninsula. We kept in touch with handwritten letters. Her words told cannery stories about characters, comedy, and conflicts. She also asked about me.

"I miss you. I miss your eyes, your voice, the feel of your skin," I wrote back, looking forward to picking up where we had left off.

The following winter, I moved into a dry, one-room log cabin without electricity and heated by a small wood stove. It was a few miles from campus, off the cross-country ski trails near the top of Miller Hill. Without a car, we spent many winter nights on campus, crowded together in Peggy's dorm room bed when subzero temperatures kept me from my commute. Our heads on her pillow, we would sometimes whisper about children and family.

"I want to have seven kids with no TV in the house," Peggy told me, our bodies touching, "and have them when we're young, so we can be young with them, too."

"I don't know about seven!" I replied. "I want to travel the world with my kids, sharing its wild places, its cultures, its tropical mountains and subtropical beaches."

With her small frame and smile that can light up a room, Peggy has always been every child's favorite adult. She seemed ideally suited for her chosen college major of elementary education. She was playful and sensitive with her niece and I could see that she'd make an attentive and loving mother.

In May and September of 1981, before and after her summer cannery employment, we made overnight trips to the distant Delta Mountains in the eastern Alaska Range and the nearby

Granite Tors, a low tundra plateau studded with craggy towers. Peggy had never camped off the road, climbed a rock, hiked a mountain. I enjoyed sharing these easy experiences with her. For gnarly adventure there was always ice climbing with Carl.

School took up my time with biology labs and math homework. Like most college students, I had little clue how a career might look after graduation, although my adviser encouraged academia as a goal: a professor, perhaps. Earning a Ph.D. fell somewhere along a vague timeline to "be a scientist." Until then, climbing would take precedence over everything—except Peggy Mayne.

THROUGH THE EARLY eighties, a climbing-related accident took the life of a good Fairbanks climber every year. But that didn't slow me down. By 1982, the Alaskan alpine style of climbing—like skiing a hundred miles to climb a steep new route—had emerged as my forte. Although summits still seduced me with the promise of alpine intensity, combining them with the wilderness below the peaks increasingly appealed to me. The landscapes in the foreground were so much richer in colorful experiences: animals and plants, rivers and forests, sounds and smells. The rock, snow, and ice of alpine routes were monochromatic. Climbing from river bottom to mountain top integrated all of wild Alaska in a satisfying tapestry of nature.

Out of food during a storm in March of 1981, three of us retreated from a stern face on a peak called Ninety-four Forty-eight. Waiting in the tent, I was restless and suggested that we ski across the tundra plains to the highway to salvage our trip. Neither tentmate was interested. I left for the road anyway, without a map, tent, or partner. It took fifty-five hours to cover the fifty-five miles along a route paralleling the glaciers that Carl and I had skied the previous year after Ten Nine Ten.

Alone on the tundra I wondered: *Which way is faster? Perhaps a race could tell.*

In Fairbanks, the idea of a ski race the length of the Hayes Range was met with curiosity, mostly at Sandvik when someone broke out the beer. The idea gained critical traction later that fall. During a guides' association meeting on the UAF campus, an impish guy in his mid-thirties arranged a stack of flyers. I picked one up and read: "Alaska Mountain and Wilderness Classic: An Overland Footrace from Hope to Homer. Carry all needed equipment and food. No roads, no pack animals, no caches, no outside assistance. Finish with what you start with."

"Hi there." The imp grinned. "I'm George Ripley. You look interested in Hope to Homer." Ripley had a round, open face with ears that stuck out as if he were really listening to whatever you said.

"Yeah, I *am* interested. I want to put on a race, too, but in the Alaska Range—a ski race from highway to highway."

George's grin grew. "Why don't you come do my race first? And then we can do your race."

That August, ten of us, Alaskans all, lined up in Hope near Anchorage. By the end of the first day, Dave Manzer, a twenty-seven-year-old Anchorage resident, had caught me. By the second evening, the Skilak River had stopped us both in our tracks. Gray water churned in a single channel seventy yards across. This would be our first river swim on the race course. Intimidated, we decided to make camp. Pulling out a candle, Manzer dripped wax on tinder and started a campfire.

"Want some tea?" he asked.

As an alpinist, I had never had much use for campfires, but welcomed this one's cheery warmth. Soon, other racers caught up to us, including a white-haired, fifty-five-year-old named

Dick Griffith. With a quiet confidence and chiseled features, he resembled Clint Eastwood in tennis shoes and a backpack.

"What are you doing here?" asked Dick, dropping his big pack. "I thought you young guys would be halfway to Homer by now!"

"We're waiting till morning to swim across. It'll come down after the sun gets off the glaciers," Manzer said.

"You gonna *swim* that?" Dick asked incredulously. "You can't swim these glacial rivers! They're too cold and fast. How you gonna swim with all that stuff on your back?"

We nodded, wondering that, too.

Dick chuckled, pulling a red Viking hat with blue horns over his head, and said, "You may be fast, but you young guys eat too much and don't know nothin'." He shook his head, the blue horns wagging in scorn.

"You need one of these," he said, reaching into his backpack to unroll a small, one-man vinyl inflatable raft at our feet. It looked like it weighed only a few pounds.

"What's that?" someone asked.

"That's my secret weapon." He chuckled again. "Old age and treachery conquer youth and skill every time."

Manzer and I looked at each other. "He's going to use that thing to float the Fox River, too," Dave whispered. The Fox River valley was twenty miles of thick alder brush and swamp that would take us more than thirty hours to cover on foot. Dick would paddle it in five.

In the morning, Dick blew up his little packraft and rowed across. Manzer tied into a rope held by a race official for safety. As he swam into the current, the line tangled dangerously in his legs. Manzer struggled and Dick rowed out to save him from drowning. It was a sobering lesson and I swam untethered. We hurried onward to warm up.

The next day we caught George, who led us along animal trails through thickets of dense brush. "Game trails are the way to go, aren't they?" he asked rhetorically, looking over his shoulder at me with his impish grin.

Manzer's campfires, Dick's packraft, and George's game trails offered me new lessons in wilderness travel. I would pass their techniques on to my son like my uncles had taught me theirs a decade before. The Wilderness Classic—as the race would come to be known over its thirty-eight-year history—would ultimately transform Alaska from inaccessible wilderness to multisport playground for me, Peggy, my son, and my friends, especially the use of Dick's "secret weapon," the packraft.

A FEW MONTHS after the Wilderness Classic, Peggy and I headed south for six months, wanting to see the world as independent travelers who find their own way. Peggy Mayne was no stranger to life on the road. Born in Massachusetts, she went to elementary school in Ohio, then Oregon. When she was twelve, her father drove his wife and six youngest kids to Alaska, where he hoped to get rich in the "bush," that part of the state beyond the road system.

The Mayne family lived first in Tok near the Canadian border, then later in Selawik, an Inupiaq village above the Arctic Circle in northwest Alaska. Peggy graduated high school during the year her parents spent in Anchorage between stints teaching in bush villages. Soon afterward, she left for UAF, happy to be away from an alcoholic and domineering father she feared.

With money we had earned from working at UAF all summer long—me in the carpenter's shop and Peggy in the paint shop—we drove to Mexico in a beat-up little red Toyota pickup that I'd bought for five hundred dollars. Along the way we gawked at the Canadian Rockies, snowshoed through Yellowstone, rock-climbed in Yosemite, and hiked across the Grand Canyon. In Arizona, we parked the pickup in Tucson, bicycle-touring across Sonora and the length of the Baja Peninsula. We spent three months in Mexico: biking, hiking, climbing, and eating. Afterward, we drove east to visit family, then north to Alaska.

Peggy often balked when I pushed her toward her limits on those adventures. Her overprotective father, worried about the safety of ten kids on a middle-class income, had never bought Peggy a bike, never taught her to swim, never set up camp away from the road, nor allowed her to take a chance that might result in injury. His strategy had worked. "Out of ten of us, nobody was ever seriously hurt," Peggy pointed out. Her rearing left her risk-averse, a complement to my risk-taking behavior.

Our experiences crisscrossing the continent taught us to communicate, share, and compromise, skills necessary to make a family work. We could also see the outlines of children more sharply in our shared lives. When we returned to Fairbanks in May 1983, my former adviser in the math department stopped me on the street to describe a new graduate offering at UAF. "Roman," he said, "this program is tailor-made for you."

Working toward an M.S. in mathematics at UAF would allow me to continue analyzing an ecological model I'd developed as an undergraduate. It would also give me financial support working as a graduate assistant and a skill in teaching math. While unscheduled on my science career timeline, the decision seemed a good one along a path that included Peggy, kids, and our future adventures together.

I had no idea that a steep mountain in the Hayes Range would hurry me on my way.

CHAPTER 4

The Cornice

Hayes Range cornices, January 1984.

Not long after we got back to Alaska, I walked into the Hayes Range to climb McGinnis Peak. At its base my partner revealed a dream he'd had the night before: I fell while leading, and to save himself, my partner unclipped to let me fall past. While only a dream, this confession against the inviolate bond of the rope unnerved me as we headed up our climb. Dangerous conditions chased us off, but I named the route anyway: "Cutthroat Couloir."

Two years later I went back with a mercurial mountaineer

named Chuck Comstock. Stocky, blond, and belligerent, Chuck was the toughest guy I would ever know. He climbed with a brutal style that many, including me, misunderstood as incompetence. On rock and ice, he thrashed like he was only marginally in control. He made hard things look desperate, scary, unnerving. He'd fall on rock, on ice, in the mountains, but somehow, he would survive to terrify—or inspire—those around him.

Like two partners at the start of a buddy movie, we didn't hit it off right away. On our first major expedition together, Chuck warned me during an argument not to turn my back, or—as he drawled in his Iowa country-boy accent—"I might sink an ice ax in the back of your head, Romin Dahl." Later, as roommates at Sandvik House, we came to blows over something petty. Cornered, Comstock landed a punch to my jaw. I replied by pummeling his belly, then throwing him on a table, breaking it and ending the fight.

Nevertheless, we partnered up for Cutthroat Couloir. We flew there in March when it was well frozen and safe from rockfall. The climb to the top took three difficult days, including my hardest lead ever on ice, a pitch we had named "Difference of Opinion." Chuck's lead on "Mixed Feelings" was even harder. After those pitches of steep rock veneered with thin, hollow ice, we finished the couloir and climbed a snowy ridge to the top.

We tented on the summit our third night. Below us the Hayes Range went dark as the sky turned indigo and Alaska's winter chill set in. The temperature reached thirty below zero and I shivered, tossing and turning in my expedition down parka and synthetic sleeping bag. Long before dawn, we woke and brewed hot drinks on our stove in the tent to warm up.

We felt good about McGinnis's Cutthroat. Maybe too good. We had just put up one of the hardest climbs in the Hayes Range. We knew that McGinnis's southeast ridge was another. Hubris sent us down that knife-edged ridgeline like happy cowboys on barebacked ponies. Then we arrived at a long stretch of cornices. Two of the most experienced alpinists of our generation had disappeared without a trace on Canada's highest mountain when one of these snowy hazards collapsed beneath them, sending the pair roped impotently to their deaths.

On Ten Nine Ten five years before, Carl had instructed me in negotiating corniced ridges: "Roman, if I break a cornice, jump off the other side, okay?" The idea that a rope stretched over the ridge between us would keep us both from falling to the glaciers below seemed iffy at best, but it would be the only way to safeguard our descent of McGinnis's southeast ridge.

AT THE END of a stressful two-hour lead, I slithered like an alpine chimney sweep down a big iced-up crack that split a craggy spire. Tied together with twin parallel ropes, we had reached the col between McGinnis and the next mountain. It was a good place to pull Chuck in on belay. Crisp blue shadows rimmed in tangerine stretched across the ridge. The sun would drop soon, sending temperatures to minus thirty again. The wind picked up.

Waiting for Chuck to join me, I looked ahead. Beyond the col, cornices clung to bare rocks like a white gyrfalcon's talons to a black fox's carcass. There was no place to camp here and no time before dark to maneuver through the tortured ridgeline coming up.

Tense and angry by the time Chuck arrived, I dumped my stress on him.

"Chuck! There's no place to camp! Why didn't you say something earlier, when you were leading?" We argued about whose fault it was that we hadn't made camp somewhere safe. He ended our spat with a whisper.

"Okay, fine, Romin Dahl," he said quietly, his jaw stiff from the cold. "Let's just split right here. I got a stove and a cookpot, you got a stove and a cookpot. We both got shovels. Just take your rope and give me mine and we'll go our separate ways." Chuck untied. He dropped the end at my feet.

Eyes as big as the yawning space around us, I stared at the naked tail of rope.

"Chuck . . . look, I was wrong. . . . You were right. It's my fault. I should have said something back there. Maybe we can camp on this col . . . Chuck, please. Tie back into the rope."

He looked away and spat a rat turd of Copenhagen snuff into the clean white snow.

"Really. Chuck, I mean it. . . . I'm sorry. It's going to be fucking cold soon. Chuck, man, please. *Tie back in.*"

His blue eyes squinted through blond lashes crusted in frost, his look like a runaway dog who's unsure if he should return to his owner. Reluctantly he retied. The moment passed and Chuck led off. The twin lines paid out as Comstock plowed a trough along the ridge crest, the broadest it had been since the summit. Half a rope length away, he pushed a four-foot aluminum stake—a picket—into the snow as protection, then probed the ridge with his ice ax. I wasn't sure why. Maybe looking for a camp, maybe testing the footing, when—in one fluid motion—he dropped from sight.

The rope yanked at my comprehension, reminding of me what Carl had said on Ten Nine Ten: I leaped free of the ridge. I tumbled into space, cartwheeling and bouncing off the snowy slopes below, everything passing in a slow-motion blur. Relaxed, without pain or even fear during a fall that felt very far, I wondered if I might die, like Peter McKeith. Or if not killed, about how I might be hurt, like others who'd fallen and survived, only to suffer with broken arms and legs, days away from help. I found myself praying: *Dear God, don't break any bones. If you must, take my life instead.*

Eventually I came to a bouncing, yo-yo-like stop, hanging from the end of the rope and alive. Dangling in the soft rime and sunshine I took stock—tools, crampons, pack, all intact and me uninjured—*my helmet? Where's my helmet?* I looked down. Accelerating toward the glacier at the base of McGinnis was an orange dot. *If the rope had broken, that would be me, careening like a rag doll.*

The rope—stretched tight over the ridge—had saved me. *What about Chuck?* With mechanical ascenders on the rope I pulled my way to the ridge crest. *Is the sturdy anchor on the other side Chuck's dead body?* On top, the rope sliced deeply into snow the cornice had left behind. A thirty-foot chunk of the ridgeline—five feet thick and fifteen feet wide—had broken free, the collapsing snow stuffing Chuck into a steep, dark couloir. Hoping to see Chuck alive, I peered down the nasty gash and spied a figure inching upward with tangled coils of rope hanging below.

"*Chuck!*" I called down. "*You all right?*"

"*Yeah!*" he yelled back. "I hurt my hand! But I'm okay!"

"Hold on, Chuck! I'll come down to you! Put in a screw!"

I pounded a picket into the hard-packed snow left behind by the cornice, then rappelled down to him. He looked all right: no blood, no deformities.

"Good God, Chuck. What happened?"

"Well, Romin Dahl," he drawled, more shaken than I'd ever seen him, "there was a strange hole in the snow and I bent over to look inside. I thought maybe we could camp in it, dig a cozy little snow cave. And then I was falling and all this snow was pushing down so hard on me I thought the rope would break! And when it stopped—well, there I was."

We rappelled a thousand feet into the night, until we found snow deep and soft enough to scrape a shallow bivy cave. It was crowded inside but we felt safe and lucky to be alive. In the morning we rappelled another thousand feet to the glacier, staggered back to camp, stepped into our skis, then cruised off the glacier and down an iced-up creek.

At the time, I had planned to return and make another climb on the east side of McGinnis. But once back in Fairbanks, the retelling of events on the southeast ridge revealed the truth, stark as the shadow Chuck fell into: the mountains don't give a goddamn about how good you are.

I loved alpinism like a junkie loves a fix, but I needed to quit cold turkey. It seemed far better to be an alpine has-been at twenty-five than a dead legend at thirty.

McGinnis marked the last time I would tie in for an alpinist's route.

CHAPTER 5

Cody Roman Dial

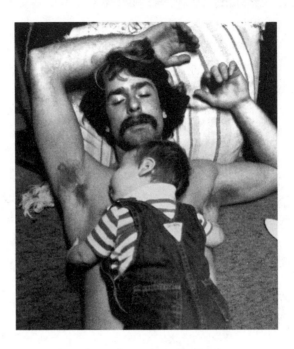

Father and son, November 1987.

After McGinnis Peak nearly killed me, I asked Peggy to marry me. We wed in June 1985 surrounded by family and friends in a wide-open field behind the Miller Hill cabin. Our honeymoon in Maui followed and then we moved into a one-bedroom house a block from Maureen and Steve. It felt good to be married and done with alpinism, although the Alaskan wilderness still tugged at my bootlaces.

In May of the following year, after finishing my master's degree, Peggy and I headed west from the Trans-Alaska Pipeline on cross-country skis to attempt the middle leg of a thousand-mile traverse of the Brooks Range. We pulled sleds filled with a tent, a packraft and paddle, plus food and equipment for four weeks, hoping to enjoy spring give way to summer. From May to August, it never gets dark in the Brooks Range.

Only five miles from the road, though, the snowpack turned rotten in the afternoon sun, leaving conditions too soft to ski, much less walk. We climbed up to the canyon rim and camped, waiting for it to cool and the snow to harden back up. "Let's just stay here," Peggy suggested when the first night hovered above freezing.

On the canyon rim, we set the tent facing east and overlooking Kuyuktuvuk Creek's valley below. We waited there, stuck in our camp, the tent hot as a greenhouse during sunny day after sunny day. We stripped off our clothes to keep cool and—technically still newlyweds—enjoyed a second honeymoon. Tent-bound, Peggy complained, "I have never eaten so much candy on one trip. We eat chocolate three times a day—just laying around in the tent!"

"What's wrong with that?" I asked, thinking it sure beat running out of food with Chuck or bivouacking on a summit with Carl.

While I was happy to lounge around naked and sweaty all day and spoon under the covers at night, Peggy wanted more exercise. The snow in the valleys remained too sloppy for travel, but the slopes above camp froze into a walkable crust in twilight. Beneath the glow of a midnight sun behind the Brooks Range's northern ramparts, we spent the week exploring the mountains each night. We would climb a few thousand feet to gain a gentle ridgeline, then slide down on our butts in a real-life Chutes and Ladders. We had one ice ax between us that Peggy used to control her descent on steep snow. I held long sturdy rocks for the same purpose.

One day, back in the tent after a night in the hills, Peggy woke, looked out, and cast her eyes up at the sky. It was three in the afternoon. "Yuk," she said, "it's still gray." The clouds meant another warm night with snow too soft for travel. "We're prisoners."

"Prisoners of love," I reminded her, pulling her back in the tent.

Chased out of the range by May's sloppy conditions, we returned to Kuyuktuvuk Creek in July to walk and packraft for the month. Midway through the 350-mile journey, we discovered Peggy was pregnant. To ease her morning sickness with something fresh, we caught grayling, a kind of arctic trout. Peggy was nervous in water, so I took her hand when we waded creeks and rivers. She sang while I paddled us both in our packraft. Each day we kept each other company; each night we kept each other warm beneath our single sleeping bag draped like a quilt across us. We fought and worked it out. And one memorable moment, we shared a deep and primal fear as a grizzly—all the while in the sights of my shaking .30-06—charged us, stopping only yards away when it finally caught our scent. Our month in the Brooks Range pulled us closer than either of us had ever been to anyone.

Years later, Peggy and I entered the Wilderness Classic, held in the Brooks Range that year. The course would head up Kuyuktuvuk Creek. In a prerace briefing, a park ranger told all us racers that the local Nunamiut name *Kuyuktuvuk* meant "place to make love many times."

Peggy smiled and glanced at me. "How did *he* know?"

NINE MONTHS AFTER our honeymoon on Kuyuktuvuk and the day a Fairbanks cold snap ended, Peggy went into labor just before midnight. "Roman, this is it. The baby's coming."

"No, it's Braxton Hicks," I said, citing the false contractions that don't signal birth. "Go back to sleep." I turned over.

She chuckled. "No, these are real. I can feel it. Let's go!" We got up and her water broke there in the bedroom. We hurried to the Fairbanks hospital in our little red Toyota. It was February 22, 1987.

Like many first-time moms, Peggy labored all night, her hair plastered to her forehead with sweat. Helpless, I could only hold her hand as she squeezed and pumped mine in rhythm with her contractions. When our newborn boy, red with blood and mucus, finally slipped out headfirst—I nearly fainted. Peggy, it was clear, was far tougher than I could ever be.

She held him and cooed, happy and exhausted like me, except I had only watched the miracle of birth: she had done its labor. I was pleased our firstborn was a son and looked forward to bonding with him as a father, and to maintaining that bond in ways my own father had not. Peggy said she wanted "another Roman," so we made it his middle name. His first name I took from Cody Pass, the wilder Alaska beyond Usibelli I had imagined as a kid. Cody Roman, I reasoned, would be what lay beyond me.

We spent the following winter in the one-bedroom house and watched infant Cody transform into a toddler. He began by balancing against walls on two feet while looking at us with awe, feeling his legs support him. His tiny hand felt good, wrapped tightly around my index finger as I walked him around the house.

One day, he sat on the floor with me and looked up expectantly. He was ten months old in a red sweater and diapers. He'd been bracing himself against the house walls for a week. I looked at him and smiled. "Stand up," I encouraged. "Stand up." Puzzled, he looked back at me. "Stand up!" I repeated with enthusiasm.

And then, shockingly, in one fluid movement, he lurched forward onto all fours, pushed off with his hands, and stood up on his own. He wobbled there, smiling as I smiled back.

"Peggy!" I called out, "Peggy! Cody just stood up! *On his own!*" Peggy ran out and we watched him take a few steps, walking and smiling with his newfound freedom.

Baby Cody slept well each night and toddled with determination. From early on, he displayed a long attention span and deep curiosity. Sometimes I carried him around in a backpack carrier on my bike or on foot. Sometimes I held him on my chest as we both fell asleep. Sometimes he cried and nothing I did would soothe him: not changing his cloth diaper, nor feeding him; not rocking or jostling him; not making silly faces or sounds. No one but Peggy could calm him then.

FALLING OIL PRICES crashed Alaska's economy in spring of 1987 with "For Sale" signs on every block in Fairbanks. When a job offer teaching math in Barrow fell through, I called an old friend, Matt, who'd taken his UAF mining engineering degree to Nome to work for Alaska Gold. The biggest gold mining company in Alaska, it could offer me a job as a manual laborer. Peggy and Cody would remain in Fairbanks while I headed west to work.

Matt, an Iditarod musher as well as an engineer, offered me his "dog shack" as a place to stay in exchange for looking after his kennel. Each morning after feeding his barking white sled dogs, I rode my mountain bike to the thaw fields, where a crew of misfit laborers melted the permafrost to mine for gold. Alaska Gold operated two gold dredges, enormous 1940s-era boats that floated in the ponds they dug. At their bow, a conveyer belt of one-ton buckets slurped up the tundra and passed the diggings to giant sluice boxes that rinsed nuggets and gold dust from pay dirt. I worked ahead of Dredge #6. My job was connecting water hoses to two-inch steel pipes sunk eighty feet down into the permafrost, then jacking and twisting the pipes with heavy tools to break them free of the ice that gripped them.

I sent nearly all my earnings home, happy to make sixteen bucks an hour. But living in the dog shack so far from Peggy and baby Cody was lonely. That spring, my high school class would hold their ten-year reunion. The invitation reminded me that my science career had been sidetracked. A decade had passed, yet I worked side by side with kids who had just graduated from high school. It was time to grow up and get that Ph.D.

After a full season in Nome's thaw fields, my visits to graduate schools on bustling campuses like Princeton and Stanford offered a contrast in cultures, and not just with Alaska's frontier. Pretentious Princeton put me off, yet the Stanford vibe was exciting. The Bay Area's nearby mountain bike trails, redwood forests, and rocky coastlines appealed to me—almost as much as Stanford's outdoorsy students and eclectic faculty.

Stanford professor Jonathan Roughgarden was a tall, lanky man who wore his mop of brown hair carefully combed to the side. Brilliant, with an owlish look befitting his Harvard education, Roughgarden beamed in excitement as he used his hands to give comprehensible shape to abstract ideas. The National Science Foundation had funded his proposal to develop mathematical models of food webs based on fieldwork with Caribbean lizards. It was a project that needed a student like me, one with athletic abilities who was also competent in the quantitative arts. From my perspective, Roughgarden and the project's fieldwork would complete my training as a modern ecologist.

It certainly wouldn't hurt to have a Ph.D. from Stanford either.

CHAPTER 6

Tropics of Cancer and Capricorn

Cody and Jazz, Culebra, 1991.

The Ph.D. program started in the fall of 1988. Peggy, who was pregnant with our second child and finishing her own degree in elementary education, stayed in Alaska with eighteen-month-old Cody. Midwinter, I flew home and drove our Subaru to California. Peggy and Cody followed soon after, flying down with frozen moose and caribou meat, part of our strategy for surviving on a grad student stipend in Silicon Valley.

A couple of weeks later we threw a barbecue potluck with my new California friends. Early on at the party and stimulated by the day's preparty cleaning, Peggy went into labor. We handed off the hosting duties to a pair of Alaskan friends and headed to the Mountain View hospital. Just after midnight on January 22, 1989, Peggy gave birth easily to our daughter. We christened her Jasper Linda Dial, her middle name honoring my mother and her first name the muscular beauty of the Canadian Rockies.

Jazzy was a sweet, beautiful baby. Her features were tiny, her smile darling on minuscule tender lips, and her sparkling personality matched both her name and arrival into the world at a party. As expected, Peggy made an attentive, loving mother of two, although a mother at times overwhelmed.

Ph.D. programs are essentially poorly paid apprenticeships with a boss expecting constant, unpaid overtime. This left me unable to pull my full weight as a parent. Alone all day in the condo complex where we lived with no other young mothers nearby, Peggy felt isolated caring for a toddler and an infant. She took a low-paying job teaching movement classes to preschoolers at a women's health club where she could bring our kids with her to work. She wasn't about to let someone else raise them.

"Roman, it doesn't make sense for me to work at the club. The kids are always getting sick there. And if I take another job, all the money I make will just go to child care. I'd rather stay home and raise them myself." Her kids paramount, Peggy gave up the companionship of coworkers to keep Cody and Jazz healthy. We agreed that emotional wealth was worth more than money, so she focused on her role as housewife and mother, an arrangement that made everyone happy.

ARTICULATING A VIABLE Ph.D. project consumed my first year at Stanford; completing it took three more. Roughgarden's NSF proposal had described the Caribbean's common anole lizards as ideal organisms to study complex food webs in tropical rain-forests. The small, colorful, and active creatures are abundant in the canopy, far above the jungle floor. At the time, scientists viewed the canopy as an inaccessible, unexplored landscape just overhead but far out of reach.

Most canopy studies cataloged life as seen through binocu-lars from a tower or the crotch of a single tree. Our research in-volved manipulative experiments in multiple trees sixty to one hundred feet above the ground. The protocol called for removal and exclusion of anoles from individual trees for nearly a year. Because anoles spend their lives in the canopy but hatch from eggs in the forest floor's soil, we would keep the lizards out of the crowns following their removal with plastic collars around the tree trunks. Next, we would compare both the counts of bugs and the amount of leaves eaten by bugs in trees without lizards to the same counts in trees with natural populations of lizards. In this way, Roughgarden and I could measure the myriad impacts of an abundant predator on its environment. That was the idea of the experiment: to bump an ecosystem and quantify its response. To make it happen would require ropework, courage, and plenty of sweat and muscle. It sounded like my kind of challenge.

Before moving to Puerto Rico, I went down to scout for living ar-rangements and brought three-year-old Cody with me. This eased Peggy's duties as she finished up with our move. It also provided my first real father-son trip with him. Together we explored a world new to us both: the tropical rainforest. We investigated giant land snails clinging to rainforest palms; watched bright green lizards do push-ups on buttressed trees; and tossed insects into the webs of hand-sized Nephila orb-weaver spiders. Cody displayed a child's innate fascination with life—biophilia, a relic from the past when children's interest in their environment made the difference be-tween life and death. Some of us never outgrow it.

As a family, we had often visited California's tide pools. Young Cody thrilled to the incredible diversity of invertebrates he found there: starfish, sea anemones, and amphipods, to name but a few. The Puerto Rican jungle offered a similar diversity, but with land creatures instead of intertidal ones. Like every three-year-old, our boy asked a bottomless barrel of questions starting with *why*: Why do lizards lose their tails? Why do birds sing? Why are flowers bright? I tried hard to nourish this insatiable curiosity on a trip that initiated our shared explorations across five continents and two decades.

Soon after Peggy and Jazz arrived, we settled into a condo a block from Luquillo Beach. With no car, we pedaled bicycles locally, pulling the kids in a bike trailer. Each morning after a mug of Puerto Rican coffee, I'd crank my mountain bike five miles and a thousand feet up into the Luquillo Mountains to work in the tree-tops. My old climbing partner Carl Tobin, himself a grad student in ecology, came for a month at the start. During January 1991, we strung horizontal traverses and vertical access lines in the forest using a mixture of mountaineering and arborist techniques I'd learned from Mike Cooper, my best friend from childhood.

Mike had started an arborist business after college. The autumn before I left for Puerto Rico to start the project, he showed me the ropes in my parents' front yard, where we went up, across, and down tall white oaks and tulip poplars. Mountain and tree climbing both use harnesses and ropes, but their use and design differ. Arborists hang from harnesses on thick-sheathed ropes meant for pulling across limbs. Their techniques for tree climbing rely on sliding ropes and clever knots, rather than hardware. Arborists move around for pay, unlike mountain climbers, who go straight up for thrills.

Mike's rope tricks allowed Carl and me to move around inside each tree's crown. With access throughout the canopy, we could mark every lizard we saw with our paint guns; it was good fun, squirting droplets of blue, pink, and yellow paint from up to twenty feet away for mark-remark statistics to estimate anole abundance. We used blue paint the first day, pink the second, and yellow the third, recording how many lizards we saw in each tree with each color scheme each day. One-color animals were seen only once; two-color twice; and three-color three times. We then applied a statistical model to calculate how many lizards we missed, based on the chances of painting we observed. Summing all these observed animals and the single "guess" gave the estimate of total lizards in a tree. We even authored a scientific paper on the arborist techniques, then unknown to the canopy science community. Among other things, we illustrated how to go from tree to tree, enabling a multiday, canopy-level forest traverse without touching the ground—a sort of "canopy trek."

Down in Luquillo, Peggy and the kids spent most days at the beach. Playing in the warm water and collecting seashells left them tanned, blond, and barefoot. Cody delighted in watching colorful reef fish through a kid-sized snorkeling mask. He stood in the shallows, bent over and holding his breath, exploring the watery world at his feet. Yards away on the beach sand, Jazz gathered foot-long tropical seedpods washed ashore by gentle waves.

Inspired by visits to my study site, Cody decided to establish his own in our yard of low shrubs and fleshy ornamentals. Marking its corners with surveyor's tape, he'd catch and release the anoles that lived there.

"Dad, I made a map of my study site!" he said, running up to me, home from my day in the jungle. He had watched me labor over my study-site map, then worked hard on his own in crayon and colored pencil. "Do you want to see it?"

"Yes! *Of course*, I want to see it!" I said, both delighted and impressed my four-year-old son had made his own map.

"Well, here are the corners. They have orange flagging." He pointed to squiggly orange Xs. "And here is where I caught a *cristatellus* in this bush." He moved his pudgy little finger to a green scribble that marked where he'd caught the brown anole with the orange dewlap and tail crest, an animal he recognized by its scientific name, *Anolis cristatellus*.

"And over here a grass anole lives by the fence. I caught him and let Jazzy hold him, too. She was careful, Dad," he assured me. Both kids knew to hold the delicate animals by a toe with the creature perched on their fist. "And here"—he moved his finger to two parallel lines—"here is where the *Ameiva* lives. He's big!" Unlike the svelte anoles that spend their time in trees or bushes, the fat-headed ameiva with its striped sides is a ground lizard that prowls the leaf litter for insects like a tiger prowls for deer: stopping, looking, moving on.

When Roughgarden heard about Cody's study site and map, he warned, "Better watch out, or he's going to be a biologist, Roman." *That doesn't sound so bad*, I thought, pleased and imagining a future when we did science together.

READING THE *WALL Street Journal* one night, I found an unbeatable airfare bargain. For the same price we'd pay to fly from San Francisco to Fairbanks, we could fly round trip from San Francisco to *Australia*. "Let's go!" exclaimed Peggy, the stay-at-home wife of a grad student pauper. A coupon clipper and smart shopper, she is always on the lookout for deals. "It's like paying for a ticket to Alaska—where we have to go anyway—and flying to Australia for free!" The frequent-flyer miles we would earn by flying to Australia and back would get us round trip from California to Fairbanks, where we went each year to maintain our Alaska residency. As Alaska residents, we were entitled to certain benefits, like no-interest student loans and the

State's Permanent Fund Dividend, an annual payout to each resident in lieu of an income tax.

With the research in Puerto Rico complete, we flew to San Francisco, left the ropes and data at Stanford, and continued westward to Sydney. From Sydney, we flew across Australia to Perth on the Indian Ocean. In Perth we rented a car to drive north to the tropics of Western Australia. Most parents would hesitate to jump into an economy-sized car for a month-long, 1,500-mile road trip with their four- and two-year-old children. But we'd had no car for almost a year. The simple novelty of being in one kept the kids entertained. Besides, there was something new and exciting to see nearly every hour in "Oz," slang for Australians' homeland.

Oz's west coast looked like California's and Baja's between Santa Cruz and Cabo San Lucas, but without the corners, cliffs, and traffic. Northward from Perth, tall eucalyptus forest gave way to Australian chaparral, savanna grassland, then desert, and finally tropical woodland. When we crossed the Tropic of Capricorn, twelve time zones from where we had lived in Puerto Rico near the Tropic of Cancer, we had come exactly halfway around the world.

We drove deeper into the Outback with its red dirt and clouds of annoying bush flies, pushing onward through the Great Sandy Desert. The desert's dunes spilled into the Indian Ocean at Eighty Mile Beach. Here we beachcombed for intricate, colorful seashells, unlike anything we'd seen before. Cody and I found a small dead pilot whale half buried in the sand. Jazz collected dried starfish and heart urchins by the dozens. Between the city of Perth and the frontier town of Broome, we watched emus and black swans, inspected road-killed kangaroos heavier than a man and with a middle toe as long as my hand; we touched curious dolphins, snorkeled over corals on Ningaloo Reef, even rode camels along a tropical beach.

Every night we tented in the Outback as a sky full of unfamiliar stars rotated into view. After dark, we cruised the roads for nocturnal wildlife. Some nights, kangaroos bounced across the pavement like basketballs on an empty court. Other nights we saw six-foot-long black-headed pythons, or caught cat-eyed geckos, once even picking up the spiny anteater called an echidna, a unique egg-laying mammal the size of a melon but poky with spines. We studied each catch in our headlamps, took its photo, then let it go safely off the road.

The next morning we would break camp and drive onward to find still more marvels: a blue-tongued skink the size of Arizona's Gila monster and looking just as venomous, with a long, royal blue tongue lolling out in threat; a thorny devil, the Australian version of America's horned toad; a glass snake, a legless lizard as long as my arm and named for its ability to lose a tail half its total length. We climbed into slot canyons carved from iron ore by flash floods and swam though their cool pools beneath white-barked fig trees whose roots grew plastered to red-rock walls. At sunset, feeding dry mulga sticks to a crackling campfire, we would watch flocks of hundreds of gallahs—crow-sized pink cockatoos—fly over our deserted desert camps. I couldn't help but reflect back on pillow talk Peggy and I had shared a decade before when we fantasized about how we would raise our family: we were living those dreams here and now in Oz.

By the time we reached Fitzroy Crossing in Australia's remote Kimberley, we'd gone feral, the kids dark in tropic tans and red dirt, their hair bleached, their bright blue eyes wild. We turned around and drove the 1,500 miles to Perth in three days, then flew home. Back on the Stanford campus, we shared our trip at slide-show potlucks, where Peggy and I listened with our friends enthralled as four-year-old Cody narrated the travelogue himself. The four of us looked forward to more family trips like our marvelous month in Oz.

THAT FALL, I began analyzing results and writing my dissertation. Expectations were high among my grad student cohort—a group that included a future winner of the MacArthur "Genius" award and others destined to be Stanford and Harvard professors. The pressure to perform was stifling. Even so, looking at my data to uncover the workings of a tropical ecosystem excited me as much as climbing a frozen waterfall unroped. Science itself, without the punishment and pettiness of peer review, still electrifies me thirty years on.

In February 1992, the year Cody turned five, a Sandvik friend called to tell us about an assistant professor position in ecology at Alaska Pacific University. After reviewing my application, the search committee invited me up. Arriving at the interview in April was a bit of an eye-opener. The brown lawns and dirty roadsides of Anchorage, littered with a winter's worth of trash, were dreadful. APU itself seemed like a ghost town, with few students in its sixties-era buildings.

Still, Alaska was where Peggy and I had always planned to settle, a place we called home, where family and old friends lived. When the search committee offered me the job, we were thrilled. APU might not have provided much in the way of pay, but we could raise Cody and Jazz in America's healthiest environment, with wild foods, clean air, and water. Best of all, we could share with our children the Alaskan wilderness just beyond Anchorage's city limits. I accepted the position and we drove north from Stanford at the end of the summer.

After my first year at APU and during our first full summer back home, I set off with our son to explore Umnak, a remote Aleutian island of geysers, glaciers, and fog. We'd moved back for just this sort of experience and I was eager to get started.

CHAPTER 7

Umnak

Cody, sixth birthday, 1993.

Hand in hand in late summer 1993, six-year-old Cody and I walked off a jet into wind-blown mist and brash gusts of rain. The wet air smelled of beached kelp and diesel. Round green hills of tall grass and broken cliffs rose above a bay lined with sheet-metal warehouses and filled with boats of all sizes. We'd landed in Dutch Harbor among the Aleutian Islands, far south of mainland Alaska. It felt warm for mid-August, when autumn lurks just around the corner in most of the rest of the state. Dutch seemed too small to be the richest fishing port on earth, where crab boats, trawlers, and other vessels deliver their catch for the world's seafood markets.

Among the three hundred islands of the Aleutian chain, I had settled on Umnak—just west of Dutch Harbor—because of its geysers and history. The ruins of Fort Glenn, a secret American military base from World War Two, sprawl across one end, while the Aleut village of Nikolski nestles in a bay on the other. Between these two sites of human habitation stretches a verdant wilderness of rolling hills, black rocks, and volcanoes with tongue-twisting names like Vsevidof and Recheshnoi. Umnak's geysers, the only ones north of Yellowstone, were the real draw, a geologic wonder I hoped to share with my son.

I wanted to walk the sixty miles from Fort Glenn to Nikolski. I'd done my homework and found the geysers on a map of Alaska's geothermal features, then phoned an old geologist friend, Roman Motyka, for information. Motyka sent me his published journal articles describing Umnak's thermal features in scientific detail. He said a single family lived at Fort Glenn and harvested the island's feral cattle. Motyka also told me about a guide named Scott Kerr who'd made Nikolski his home. After talking to a half dozen people familiar with Umnak and poring over maps of the island, I sketched out a route suitable for a soon-to-be first-grader.

From Fort Glenn's airfield we would head west along the Pacific coast, then hop over the island to the geyser basin on the Bering Sea side, then back again to follow the bases of Recheshnoi and Vsevidof along the Pacific. The ocean side, peppered with black beaches on a ragged coastline, would offer tide pools, Cody's favored habitat for exploration and discovery. The island also felt safe from Alaska's biggest hazards: Umnak has neither bears nor large glacial rivers.

But danger did exist there. Separating the chilly Bering Sea from the warm Pacific Ocean, the Aleutians suffer the worst weather in the world. Always windy, often rainy, mostly foggy, the archipelago is known as the birthplace of storms. While winters rarely see subzero cold, summers are cool and cloudy. Like mountaintops above the tree line, the Aleutians support no trees or shrubs over knee high.

With hypothermia a very real threat, especially for a little boy, Umnak's weather worried me. A one-piece Gore-Tex suit over long underwear and fleece pants and sweater would seal him in from the incessant wet wind. Pulling on his orange rain pants and jacket would protect him from a driving rain. I would fuel him with his favorite snacks kept handy all day, then quickly have him change into dry clothes each night for warm sleep. Our dome-style mountain tent would shelter us from gale-force wind and rain. And by tucking a copy of *Charlotte's Web* into my pack to read aloud before bed, we could bring a little of home with us to the wild.

Full protection from Umnak's weather was key, but the remoteness itself between Nikolski and Fort Glenn was a risk. Remoteness was not unfamiliar to us. We had driven for days across the Australian Outback when we would see few other cars. As a family, we'd day-hiked in the front country and backpacked for two and three days in the backcountry, including trips with grizzly bears and glacier river crossings. Keeping Cody safe would be simpler without bears or big rivers, but we would need to avoid accidents with a careful, cautious route choice.

Peggy encouraged our journey. She knew firsthand how time spent together in the wilderness strengthens bonds and relationships. And she knew that I would be sensitive to Cody's needs and fears—looking after him, keeping him safe. But she also voiced her concern: "What if something happens to you?"

My answer begged her question: "Peggy, what could happen? I'll be careful."

"You said there are wild cattle. I don't want you guys unprotected if a bull decides to charge. You should take a gun." I packed a .44 Magnum.

The responsibility to keep both Cody and me safe from hypothermia, drowning, animal attacks, and injuries went without saying. But beyond safety, I wanted this trip to initiate a lifetime of shared wilderness adventures. For that, Cody needed a profound experience that he would want to repeat. Like most parents, Peggy and I replicated the positive aspects of our own parents' child-rearing, tried to avoid the negative, and defaulted to the rest. If I wanted Cody to join me on future wild trips, then I needed to notice what interested him.

FROM THE DUTCH Harbor airport, Cody and I hopped in a taxi and met up with George Ripley, organizer of the first Wilderness Classic. At George's house, out of the wind and rain, we talked about our trip. Our pilot, Tom Madsen, had been flying a big Japanese group of mountain climbers and their camera crew up and down the chain all summer. Every island hop needed multiple flights and there would be an empty seat the next morning. Madsen could drop us off at Fort Glenn on his way to the Islands of Four Mountains, just beyond Nikolski.

In the hangar the next day, the Aleutian guide Scott Kerr helped the Japanese team load up. We'd spoken on the phone but never met. He turned away from a mountain of gear to shake and hand me a six-inch aluminum tube, a fraction of an inch across.

"What's this?" I asked.

"A tent-pole splint. I've been with the Japanese for over a month and they've climbed five volcanoes from Unimak to

Umnak. But man-o'-man, the weather can get pretty fierce out here. The wind broke poles in three of their tents. Take this, for when the wind breaks yours."

Cody and I followed the Japanese into Madsen's brown twin-engine Beechcraft and crawled into a back seat. A stack of duffel bags and boxes stuffed in the tail crowded my shoulders and head. Cody sat on my lap with the seat belt buckled across us both.

Madsen bounced the plane off the tarmac and into the wind, headed for Umnak. A stormy low-pressure system would move in for the next couple of days, he warned, then it would clear up for five more. The news of good weather to come relieved me. After a short, bumpy flight, we circled Fort Glenn's airfield. The ruins of the old military base stretched across the coastal tundra near Okmok Caldera. Madsen made a tight bank into the wind, then dropped and landed on the mile-long cinder runway. Cody and I hopped out. Madsen wrestled our single sixty-pound pack from behind the seats.

The wind blew stronger here than in Dutch. None of Fort Glenn's original structures remained intact, save those whose four corners were anchored to the tundra by cables. The rest—walls, roofs, floors—had been gutted and scattered by Umnak's incessant, erosive winds.

A sturdy guy in his mid-fifties, along with his wife and adult son, rode out on ATVs to meet the plane. Madsen exchanged mail and greetings with Fort Glenn's only residents, then hustled off to fly his Japanese passengers to their next destination.

As the plane taxied away, I stepped forward. "Hi there, I'm Roman Dial."

I hoped the guy had heard of me, maybe from an article about the Wilderness Classic in the Anchorage newspaper or *Alaska* magazine. Name recognition can help outlandish plans seem reasonable to strangers, but his skeptical look made it clear he wondered what the hell a man with a little boy was doing on a remote Aleutian island.

"I'm Gene Maynard. This is my wife, Rene, and my son, Cloud." I shook hands with his family while Gene bent down to Cody.

"And what's your name, little guy?"

"I'm Roman Two," he replied, grinning.

What? I thought, startled. Up to that moment, he had always called himself Cody. I smiled broadly, choked down a giggle, wiped an errant tear from the wind.

"Roman and Roman," laughed Maynard. "Well, what do you know! Come on up to the house, Roman One and Roman Two."

From that moment on, Cody Roman Dial would introduce himself as "Roman," a name Peggy, Jazzy, and I would address him by, too. Female relatives—grandmothers, aunts, and cousins—would continue to call him "Cody," while my dad would call him "R2" in an affectionate attempt to differentiate us. At home, Peggy addressed "her two Romans" with a subtle yet unmistakable difference in intonation.

"Jeez, what ya got in here?" Gene wheezed, throwing my pack on his three-wheeler. "Hop on."

Gene motored us to the ranch house, one of three intact structures at Fort Glenn. The other two were his sheds next door. An inch-and-a-half-thick cable anchored his house to the turf. Inside, their place was small and cluttered, like most cabins off the road system in bush Alaska—like my grandmother's farmhouse, for that matter.

"So." He looked me square in the eye. "What are you doing out here? Hiking around?"

"Yeah, I guess." Gene and his family would be our first source of help should something go wrong. He needed to know our plans. "I want to walk to Nikolski. With my son."

"Nikolski? That's more than fifty miles away. You sure you wanna do that?" He looked down at Cody Roman. "That's a tough crossing. I don't know anybody who's made it all the way."

The skepticism in his voice rang all too familiar: like Dieter in Yosemite or Alaskan boaters hearing about packrafts. Once, a mountaineer even bet me and my Olympic-caliber partner a thousand dollars we couldn't ski the length of the Hayes Range in less than a week. We finished in three days.

I deflected the conversation. "How about you? What are you doing out here?"

"Oh, we're runnin' a cattle business. There's a couple thousand head of cattle brought out here after World War Two. All we gotta do is get the beef off the island." He frowned as a gust of wind shook the house. "That's the tough part."

"How long you been at it?"

"'Bout six years now. But thinkin' to get out."

He circled back. "You know, there's a couple o' cowboys here who took horses out toward Nikolski. But they couldn't cross the river. Had to come back."

"River crossing, huh? I've crossed some rivers," I said, feeling compelled to display some credentials, to tell him about swimming the Skilak or shouldering mountain bikes across a dozen rivers bigger than any between Umnak and Adak. But I knew this game well, mostly from losing. The more experience I claimed, the more desperate for acceptance I sounded, the smugger he would feel as a local. I kept my mouth shut.

Gene harrumphed. "I'll go get those guys—you oughta talk to 'em anyway." Gene opened the door to leave and the wind charged in like a junkyard dog. Rene shouldered the door closed behind him.

As we waited, I studied a framed photo on the wall of a bronco rider with an arm outstretched and bandana flying. The cowboy had been caught midair in a packed stadium. The horse's hooves looked six feet off the ground. "That's Gene," Rene volunteered. "He rode in the rodeo." Gene Maynard had been a saddle bronc–riding champion for much of the sixties and seventies. Clearly, he, too, had balanced physical risk against emotional reward.

Gene came back with two guys around my age. They sounded Canadian.

"So, you guys rode over toward Nikolski?" I asked, engaging them in trip-sharing talk.

"Yeah. But we couldn't make it. The river was too deep." Their lips tightened. "You sure you want to try that with the little guy?" These cowboys were lean, fit even, but they didn't look like the wilderness jocks who raced in the Classic and confronted crossing after crossing. *Do they even know how to read rivers, how to pick a good crossing or time of day?*

"Which creek was it?" I deferred, holding out my map, probing for their depth of Umnak knowledge.

"Just before Amos Bay. It's a long ways back to here if you can't get across. How much food you got?"

"Eight days, ten if we stretch it. Which river exactly?" The information would be valuable.

They looked at the map. One of them smeared a finger around the southeast slope of Recheshnoi where its biggest glacier fed a four-mile stream to the sea. "Here."

A few days of incessant rain or an afternoon of sun could swell a creek like that into an uncrossable river. "Hmm. By the coast?"

"Yeah."

"Well, if we get there and it's too high, we'll camp. Wait for morning when it's lower or for the rain to let up if it's raining. The forecast calls for good weather after a couple days."

"How old's your boy?"

"I'm six," Cody Roman piped up.

Now they worked him.

"What do you think?"

"I think it's windy," the youngster responded seriously.

The river crossing was just a proxy for their main, unspoken worry, the same as Peggy's and the same as mine. *What if something happens to me and my little boy is left on his own? How responsible is that?*

CHAPTER 8

Space Captain

Umnak Geyser Basin camp, 1993.

The young, newly minted Roman and I left for Nikolski. I was alone with my six-year-old on an empty Aleutian island, and doubt gnawed deeper: *What if something happens to me?*

Soft green tundra stretched to the base of Okmok, where clouds wrapped around black lava towers like washrags around bad plumbing. Flowers familiar from the hills above Anchorage—Indian paintbrush, monkshood, and harebells—even the swards of grass—all seemed pumped up in size, as if the Aleutian wind had inflated them.

We made camp early, before the rain soaked us, taking refuge in our roomy yellow tent. A decade of sunshine on glaciers, deserts, and tropical beaches had faded the tent fly from blue to gray. The cowboys' questions and Scott Kerr's story of tents torn open left the shelter looking inadequate as protection against bad weather.

Madsen's promised storm arrived with nightfall. Its gusts came in powerful waves with unceasing rain. Three times hurricane-force blasts barreled down Okmok, rumbling like a locomotive before they hit. Each gust crashed into us, collapsing the tent and plastering wet fabric on my face. But after each blast, the tent miraculously sprang upright.

In my sleeping bag, eyes wide in the dark, terrified that a pole might break and rip the nylon fabric to leave us exposed to hypothermia, I asked myself: *Why the hell did I bring him here? What kind of dad am I really?*

Somehow, Roman slept through it all and the tent remained intact.

The next day, clear skies revealed a peculiar landscape of corduroy-textured green domes and coarse black cliffs. Tall grass reached to Roman's waist and pulsed in traveling waves up and over the summits of the surrounding hills. Bundled in his one-piece Patagonia suit with its hood pulled over his hat, he marched onward, waving his red mittens in time with his steps.

We passed a herd of forty piebald reindeer milling about on a low divide, then descended to camp at Hot Springs Cove on a Bering Sea beach. Grass crept up and over dunes of black sand. Waterfalls spilled from high cliffs, but the wind blew their discharge upward against the laws of gravity.

Roman wanted a fire. It would be tough to get it going, but what kind of outdoor father would I be not to start one? With dry grass, driftwood, and persistence, we built a campfire behind the dunes, pressing ourselves together and next to its dry warmth.

My boy poked at the burning driftwood with a stick, keeping the flames alive and cheery.

"What's fire, Dad?" he asked.

I thought for a minute, searching for truth in simplicity. "Trees make wood by gluing parts of air and water together with sunshine. When the wood burns, the sunshine comes back out as fire and the water and air go up in smoke."

He looked at my face for some hint of jest, then turned to study the coals.

"Is that why fire makes light? It's sunshine?"

"Yep. And the water comes out as steam. That makes the smoke gray."

The night passed calm and clear, the morning hot. We dried our gear and relaxed on the warm black sand.

By the third day of hiking, Roman didn't complain about sore feet or tired legs. Bundled in his pile clothes and one-piece suit, he trod along, looking for strawberries, blueberries, and sweet nagoon berries to pop in his mouth. He'd pick up interesting rocks and hollow grass stems that he called straws and piped between his lips. I felt a parental profoundness in simply watching him engage so purely as a child with his creative attraction to nature.

The weather held—windy, but never cold or wet.

Climbing out of Hot Springs Cove was steep, yet Roman managed it well. Our family day hikes up gentle mountains near Anchorage had prepared him for climbs like this. As we descended the other side, we could see puffs of steam rising near Geyser Bight Creek.

"Look at that, Roman!" I called, trying out his new name.

A miniature Yellowstone, the geyser basin was sized just right for a six-year-old. Knee-high geysers gushed over limey aprons, their discharge spilling as hot little waterfalls into the creek.

Fumaroles roared, mud pots plopped. Even here, five miles from either coast, we could hear the ocean waves crashing. Recheshnoi, draped in small glaciers and broad snowfields, rose above the marshy valley.

One hot spring—a deep indigo at the bottom with a pool of blue rimmed in a rainbow of green, yellow, and orange—held a pile of reindeer bones.

"What happened to the reindeer, Dad?"

"He probably got too close in the wintertime and fell in," I guessed, thinking back to bison bones in Yellowstone.

"Why would he get close in winter? To stay warm?"

"Maybe, or maybe because the snow was deep everywhere else except here."

I told Roman that people name hot springs and geysers. He dubbed the hot-spring "Caribou Stew," chanting the rhyme as he tossed in a rock.

Wet meadows filled the valley between thermal features. We camped on a low spongy dome where water oozed through the tent floor, warmed by the basin's thermal activity. "Feel this, Roman," I offered, my hand pressed against the tent floor.

His eyes lit up. "It's warm."

Dry in sleep clothes, we scrunched together on our overlapping foam pads. Perched there, I read three chapters of *Charlotte's Web* aloud. Roman studied Garth Williams's illustrations and searched the text for words he recognized. We reclined on our sleeping bags and pushed ourselves close to share the book and our love.

On the fourth day, we walked hand in hand up a Roman-sized babbling brook that splashed over rough, hardened black lava. He asked questions that six-year-olds ask to make sense of their ever-expanding world. He made analogies involving Legos and reminisced more about Jazz than about Peggy. Apparently, one parent could replace the other, but a parent couldn't replace a sibling. He talked about kids at school, like Vincent Brady, who would be a lifelong friend, and the things that they'd done together.

As we neared a pass that led back to the Pacific side, the landscape went lunar. We were both taken by the otherworldliness of the place, with its rugged black rocks, sand and gravel, the total lack of plant life. The scene ignited our imaginations and we slipped into role playing as explorers on another world.

"Captain," I asked, inspired by the barrenness, "where are we?"

"On another planet," he answered, jumping into the game without pause.

"Be careful, Captain," I went on, encouraging him. "There might be monsters here."

"Who are you?" he asked.

"I'm your sergeant. You command me."

There were space hazards everywhere. My pack carried oxygen that we shared when the Captain ran low as he battled aliens. Asteroids fell around us but our wind-shells-turned-space-suits protected us in an extraterrestrial universe. Roman led us onward, incorporating the ground we covered in our role-playing fantasy.

Two hours off Earth went by quickly, easing the difficult terrain.

Once over the pass, we stumbled across a long stretch of rubbly 'a'a, the Hawaiians' word for sharp and broken, clinker-like lava. Whenever it was especially rough, Roman slipped into space captain mode, his imagination lightening his challenges. Eventually, small patches of plant life reappeared, then green carpets of heather and dwarf willows. We walked along tiny beaded streams that snaked through low-growing gardens of wildflowers. There were occasional shallow caves, too, at the base of short lava cliffs.

A bit nervously, Cody Roman peered into a stream that disappeared as it went subterranean.

"Pretty cool, huh?" I asked.

"Nah, I don't like it. It's dangerous," Space Captain said. "C'mon, Sergeant."

Arriving at the Pacific beaches relieved me. Injuring myself was my greatest fear. Falling on the 'a'a and breaking a leg seemed a real possibility as I labored under a fifty-pound load of food, camping gear, and clothes for us both. In order to keep the trip more enjoyable for him, young Roman carried no pack.

Sometimes, when Roman's feet hurt, I put him on my shoulders, doubling the weight that I carried. There'd been very few carries that day, thanks largely to the Captain's active explorations and battles with aliens. It had been our longest, hardest, and highest day yet. The setting sun cast long shadows and a warm glow across Recheshnoi.

The next day broke sunny, without a breath of wind, but so many mosquitoes! We could see why. The cattle were as thick as the bugs. We hiked along the coast to distance ourselves from the lone bulls who tore at the dirt with their hooves and threw dust over their backs with long pointy horns. I kept the .44 handy.

Sea lions croaked and puffins dove beyond the surf line. Alongshore, Roman delighted in feeding the green tentacles of sea anemones with small creatures he caught and named, like "dinglehoppers" and "jumping jacks." He wanted to catch a salmon, but we saw none in the streams, including the knee-deep, glacier-fed river that had repulsed the cowboys. We camped near it to get an early crossing when it would be lowest.

The cowboys must have been here after a big rain. I crossed it first with my pack; then carried Roman. It was easy.

Along the coast we walked over lava reefs, uncovered by the low tide and teeming with life. Isopods—ancient, trilobite-looking relatives of shrimp the size of my thumb—crawled so slowly across blades of brown kelp that they hardly seemed to move. Years later in graduate school, Roman would study the geography of these creatures' genetics—a sign, perhaps, of how deeply our first "big trip" had touched him.

WE WERE CAMPED fifteen miles from Nikolski. Mist arrived with morning, burying Umnak in fog. We broke camp in wind and rain on our last and longest day of the walk. I pulled out the compass, showing Roman how to read it. "Press the compass to your chest and keep it level," I instructed, looping its lanyard over his head. "Now turn your body to put the red arrow in the red shed and when it's in there, keep it there by walking straight."

The day passed gray and dull, with visibility limited to fifty yards and fewer. The wind whipped our faces with the draw cords from our rain jacket hoods. Cody Roman worked his imagination, naming the hills we climbed, names I didn't record in my journal with so little time to write each night. His feet hurt more in bad weather than good, so I took to longer carries. "Son," I asked at one point, "can you walk a while? My back hurts."

"Sure, Dad." He slipped off my shoulders, lightening my load by half, taking my hand as we walked side by side. An hour or two later, out ahead leading the way with the compass, he stopped and asked, "Dad, can you carry me? My feet hurt."

"Sure, Rome, let's take a break first." Clothes sticky with sweat under rain gear, we collapsed on my pack. Resting there, sheltered from the wind by tall grass, he chewed on a strip of jerky, then unwrapped a yellow Starburst candy, his favorite.

When we finally reached Nikolski, the weather broke. The Islands of Four Mountains—green, perfect cones striped in snow—rose from a deep dark sea like art on a Japanese scroll. Whale bones and skiffs rested outside the homes of Nikolski's three dozen villagers. Orange fishing floats hung under the eaves of their weathered wood houses. A hundred yards up the beach from the surf, bleached white logs traced the power and reach of winter storms. There weren't any forests for a thousand miles.

We found Scott Kerr, the guide who'd given us the tent-pole splint. In his warm little house, we felt free from the elements at last.

News spreads quickly in rural Alaskan villages, especially about six-year-old boys who walk sixty miles. A weathered old Aleut in a Carhartt jacket named Simeon Peter Pletnikoff dropped by Scott's place. During World War Two, nearly all Aleuts, or Unangan as they call themselves, had been forcibly removed by the U.S. government. "Aleut Pete" had been allowed to stay and help fight because he was such a talented outdoorsman who knew the Aleutians well. As one of a motley army squad known as Castner's Cutthroats, he fought against the Japanese soldiers who had invaded his island chain.

Aleut Pete sat down and cupped his hands around a mug of coffee. Round glasses on high cheekbones suggested a gentle wisdom. He raised his eyebrows, smiling. "Well, aren't you a strong little fella? Not many people who aren't Unangan have walked across Umnak. Were you scared?"

"Sometimes," Roman admitted, "like when the wind blew our tent down. Or when the bulls were ripping up the dirt with their feet. Or when the geysers went too big. But look what I found!"

Roman opened his hand, revealing an orange-colored agate. "A volcano made this!"

In his other hand, he held a polished black rock, smooth and round as a ball bearing. "And this one, too. I found it by the river where we camped. And my dad found a glass ball in the seaweed! And big, big barnacles that came from a whale's back!"

Aleut Pete sat back, his smile widening at the young boy's enthusiasm. Then he reached into the pockets of his Carhartt jacket and pulled out a glass ball from a Japanese fishing boat and a sea lion's canine.

"Here you go, little hiker. Something to remember from Umnak."

After the visit with Aleut Pete, Kerr led us to Nikolski's deserted school and let us in. We spread our things on the carpeted floor and felt grateful to sleep someplace other than a damp, flapping tent. I hoped the scheduled flight from Dutch would get us straightaway so Roman could return to his sister and mom. Instead, as often happens in bush Alaska, we waited for the weather to clear. We waited a week.

The first night in the school, cozy in our sleeping bags, I felt pleased with my young son's performance over the previous week. There'd been no complaining from the curious and imaginative six-year-old, asked to walk all day, every day, for a week. "Roman, you're a good hiker," I praised, "and a strong one, too. Jazzy and Mom are going to be very impressed with you."

We had averaged one mile an hour, eight hours a day, making twelve and fifteen miles the last two days. But the numbers didn't matter. We'd grown close. Roman had learned about nature and about himself, how to deal with discomfort, wind, and rain, walking day after day. I'd learned how to pace, care, and sacrifice for my son.

Maybe it was too soon, but I asked anyway. "Roman, did you like this six-day trip? Would you want to do one again?"

"Yeah, Dad, it was okay. But let's bring Mom and Jazzy next time."

"Okay. We will," I promised, smiling at the thought of a Dial family adventure.

"Now. Can we finish *Charlotte's Web?*" he asked.

We pushed ourselves together to read the final chapters and fell asleep contented, trip partners for life.

CHAPTER 9

Borneo

Draco, Malaysian Borneo, 1995.

In the heat and humidity of the equatorial night, Peggy and I lay naked, unable to touch. Outside our open hut, tree frogs piped incessantly while katydids sawed and cicadas buzzed, a cacophony pierced by the screeches and hoots from an inky blackness. We were careful not to press against the flimsy mesh of the tent-like mosquito net that protected us from Borneo's biting insects and the diseases they might carry. The kids, now six and eight, slumbered under their own mosquito net in a bed next to ours.

We were in Bako National Park on Asia's largest island. Bako was meant to be a warm-up for Gunung Palung National Park, or "GP," where we planned to spend a month. GP is a roadless Indonesian wilderness of rainforests, mountains, swamps, and rivers. Its sole structures then were a small collection of tin-roofed open-air huts set deep in the jungle, reached only by dugout canoe. The primitive Cabang Panti ("Cha-bong Pon-tee") Research Station served as base camp for the scientist or two working there at any given time. A network of trails and incredible wildlife, not yet discovered by *National Geographic,* provided a glimpse of tropical rainforest perhaps unmatched in the world. I had visited GP the year before and left a changed man.

"Indonesia is a lot more primitive than this," I told Peggy. "Nobody speaks English. There's malaria, dengue, hookworm. I'm not sure we should go, really. I'm worried about the kids."

Peggy turned to me. "We've come this far. We're close. Your pictures from last year make it look amazing. We can protect the kids. And you know where to go and how to get there. I think we should do it."

TROPICAL RAINFORESTS IN Asia, Africa, and South America have long fascinated scientists and laymen alike with their stunning, overwhelming array of life. Straddling the equator, Borneo's tropical rainforest supports a higher-order biodiversity than any other on Earth. As in the Amazon, there is a dizzying array of small and tiny fantastical creatures that fill every square inch of rich, green plant life. But Borneo's rainforests are twice as tall as those in South America: Borneo's dipterocarp trees grow as high as redwoods. The world's largest flower, *Rafflesia*, three feet across and smelling of rotten meat, lives there, too. Lianas— woody vines as big around as pine trees—hang from buttressed trees. Gourd-shaped carnivorous pitcher plants called nepenthes grow in spectacular diversity, with some specializing in catching bird droppings, some trapping rats and frogs, and others, with a more prosaic diet, feeding simply on ants and flies.

Unlike South American jungles, where there are few large animals of any kind, Borneo has pygmy forest elephants, dwarf rhinos, even wild forest cattle called banteng, and like the Amazon, it has big cats and small. But while the New World tropics have only the familiar white-tailed deer, Borneo has five species, ranging from the rabbit-sized mouse deer to the elk-sized sambar. The strange, fanged muntjac deer even barks like a dog. Besides eight varieties of monkeys, there are primitive primates, too, including the fist-sized tarsier that snatches insects with its alien-looking hands and the nocturnal slow loris, a small bear-shaped fruit eater. Best known are Borneo's lesser and greater apes: the acrobatic gibbon and the 150-pound orangutan. Less famous are its dozen kinds of "flying" squirrels, flying lemurs, flying lizards, flying frogs, and even flying snakes, all of which glide from tree to tree. These wonders and more live on an island half the size of Alaska. Visiting Borneo is like going to another planet.

My first trip followed an invitation from Tim Laman, then a graduate student at Harvard. We had met at an international canopy conference and with mutual interests in science, adventure, and documentary photography we hit it off immediately. A tall, mustachioed redhead, Tim studied strangler figs in Borneo's forest canopy and when he suggested we climb strangler figs together at his research site in GP, I jumped at the chance. Tim faxed me directions from a faded Xerox copy. It took me ten days in December to get from Anchorage to Cabang Panti by way of a malarial village called Teluk Melano. From Melano, I hired two locals to paddle a dugout sampan canoe to meet Tim at GP. As two young canopy scientists at the beginning of our careers, we climbed trees, took photos, and recorded observations in the forests, mountains, and swamps. We woke every morning to the delightful whoops of serenading gibbon families. We watched orangutans hang upside down by their hand-like feet feeding for hours on wild durian, then we tried the delicious fruit ourselves. And of course, we picked hundreds of terrestrial leeches from our clothes and sometimes—plump with blood—from our skin. While a nuisance, those bloodsuckers didn't deter us from going out day after day in search of wonders.

That trip and the time I spent with Tim provided the most dazzling tropical experience of my life. The entire forest was fruiting—hundreds of species of trees, lianas, and herbs—during an infrequent and irregular mast event, alive with more different kinds of vertebrates and invertebrates than I'd ever seen anywhere. A highlight was a night spent one hundred and eighty feet up in a fruiting dipterocarp, my hammock suspended over a crow's-nest-like orchid epiphyte encircling the trunk and fully ten feet across with dozens of hand-sized blossoms pollinated by thumb-sized bumblebees. That night, dew fell but no rain. At dawn, the rising sun melted away diaphanous mist clinging to rainforest giants. Awakened by the lion-like roar of a big male orangutan, I knew that Peggy and the kids had to experience Borneo.

As expected, they found Borneo's wildlife in Bako fascinating. A long-nosed proboscis monkey, the size and color of a small deer, picked leaves in the crown of an oak-like mangrove. Below him, a foot-long mudskipper crawled across the muck with a mouthful of water in its bulbous head as a sort of reverse scuba. Literally a fish out of water, the mudskipper pulled itself along using fins as legs, looking like a primitive Permian tetrapod. Back at our hut, an eight-foot serpentine dinosaur of a monitor lizard prowled the premises, its long blue forked tongue tasting the air just steps from the tables and chairs where we played Yahtzee after lunch during afternoon rains.

The kids dutifully recorded these wonders in their journals. Roman had started in Singapore, exclaiming that "Chewing gum is illeagal!" In Kuching, capital of Sarawak, Malaysia, he tasted durian and mangosteen, the king and queen of fruits in Alfred Russel Wallace's *The Malay Archipelago*. He recounted his divergent reactions in tidy print:

> I ate a mangosteen and tried something worse than brustle sprouts! The Durien! Yuck! The mangosteen looks like a giant crow berrie. You would have to sqwash it to opan up! It's the best fruit I ever tasted. It tasted like orange or yellow starburst with a tang.

Besides novel tastes and sights, we caught and handled new fauna and flora. At the edge of a city park lawn, Jazzy spotted a lone draco in a tree. Like the anoles we knew from Puerto Rico, the draco (Latin for "dragon") is an arboreal lizard that does push-ups and fans its colorful dewlap to challenge territorial rivals. But unlike anoles—and more like dragons—dracos have wings and can *glide*. Picking up a clod of dirt, I threw it at the brown lizard, knocking it to the park lawn.

I hurried over to pick up the dazed, uninjured animal. The four of us inspected a thin, delicate lizard whose back was the color and pattern of lichen-covered tree bark. He had a short nose and alert eyes that watched us calmly but warily. We gently unfolded his wings that stretched across six ribs on each side, finding bright blue and black patches on their undersides. Used for gliding, the fragile patagia were as wide as the lizard's body was long, filling the space between his front and hind limbs. We all delighted in holding such an exotic and unlikely creature: a flying lizard.

As an experiment in animal behavior, we decided to watch him glide. I gently tossed the draco ten feet into the air. At the peak of the toss, the lizard opened his patagia and deftly glided to the lawn twenty feet away.

We all looked at each other. "*Wow!*" both kids called out as they ran over to the lizard waiting in the grass.

Roman picked him up and tossed him again. At apogee, the lizard spread his wings, coasting to the lawn like a paper airplane.

Roman laughed in delight and turned to me. He flashed his teeth in an excited smile: "*Cool!*"

"Oh, Dad," Jazzy said, "that's mean! Let him go."

"Here, Jazzy, why don't you let him go. You found him. Just toss him up and toward his tree so he can glide home."

Jazzy, too, tossed him gently upward, this time toward the tree where she had first found him. The draco curved his glide toward the trunk, then nosed up on his approach and stalled, landing abruptly to scamper around to the backside of the tree where he hid from the human family that had no doubt terrified him.

Later, we would handle harmless ants the size of Jazzy's thumb and a thrumming cicada the size of her tiny fist, the bug's proboscis as long as her pinky. We would catch and release fish we had seen before only in freshwater aquaria; feed beetles to nepenthes; feel the pinch of a neon blue-fiddler crab; pull open an ant plant and watch its protectors scurry; taste a dozen new fruits and cuisines from three nations. The discomforts of heat, humidity, and the odd pockets of foul odor were all erased by the hands-on discoveries of new sights, scents, and sounds.

Eager for more, we headed deeper into the heart of Borneo.

CHAPTER 10

Gunung Palung

Snorkeling, GP, 1996.

We arrived in Kalimantan the day after Peggy convinced me we should push on to GP. Cars were absent, motorbikes few, and bicycles the most common wheeled vehicle. Up to three people at a time rode on a single bike, but most simply walked in flip-flops or bare feet. Traveling as a family, we found that people were eager to please, quick to help, happy to interact. But nobody spoke English in a countryside where houses were little more than palm-thatched roofs over half-walls on stilts.

After crossing the hundred-mile-wide Kapuas River delta by

riverboat, we waited alongside a dirt road for a minivan. Our boy and girl sat patiently on our duffels. By the time the van arrived, fifty people surrounded the kids. Caucasians were rare and their children exceptionally so. At first the two enjoyed the attention, but it got old fast. People simply couldn't keep their hands off the blond-haired, blue-eyed little children.

Reaching Teluk Melano, we stayed in a guesthouse next to the river. The *Anopheles* mosquitoes that carry malaria emerge after dark and last until dawn, so at night we covered ourselves in insect repellent and long sleeves, then climbed under our mosquito nets for the next twelve hours. Peggy tracked down and killed every mosquito that made it inside. Taking a weekly dose of antimalarials, I could duck out from under the net if needed. The kids and Peggy didn't risk the drug's neurological side effects. Once we reached the wilderness of GP, there'd be little chance of contracting tropical diseases because mosquitoes generally only carry parasites from humans who have them. Rural areas are the most dangerous; wilderness and urban areas less so.

There was no running water in Melano. Instead we took *mondis* by ladling cool water from a rain barrel to wash the sweat off our sticky bodies. At the equator, the mondi is the most refreshing way to stay comfortable in the oppressive heat.

All the way from Harvard, Tim had arranged our transport to the little village at the edge of the forest, downstream from Cabang Panti. There, in the last settlement before the park boundary, two sinewy men with muscles like knotted hardwood reached to the bottom of the clearwater stream and pulled up a sunken sampan to bail. Peggy looked at me, her eyes wide, her smile gone.

"They keep it underwater to prevent the wood from splitting. The boat will float," I said reassuringly.

While the boatmen emptied the long, narrow dugout, our kids

drew the usual crowd. These villagers, though, shared with us their local red rambutans, a sweet, spiky-skinned fruit Roman and Jazz loved to eat. We loaded our gear into a cargo boat with two more boatmen, got in with our paddlers, and shoved off. The boat was tippy with little freeboard and the seats were hardwood planks. Too afraid to move for fear of dumping ourselves into the water, we sat still for hours as our rears went numb.

At first the river meandered past spiny, palm-looking pandanus plants whose twisted stalks emerged on stilt roots from deep, black water. We weren't alone here. Gnarly men in loincloths and tattered white shirts poled skinny rafts of logs tied together with rattan.

Logging in Borneo would reach its peak in the late nineties. During our visits over the next fifteen years, the seemingly inexhaustible forests of Borneo would disappear, just as the buffalo of the American West had gone nearly extinct a century before. Instead of watching cattle replace buffalo, we would witness oil palm plantations replace rainforest.

Even GP's national park status, with its Ivy League research station at Cabang Panti and its exposure in *National Geographic* magazine, wouldn't be enough to keep the loggers out. In the 2000s, when most of the big timber outside of Kalimantan's parklands was gone, Indonesian military leaders financed locals with chain saws to cut and sell GP's giant dipterocarps. Documentary filmmakers visiting Cabang Panti to record its orangutans were forced to bribe the loggers to silence their chain saws during filming.

But in 1995 the virgin forest was empty of the sounds of motors and chain saws. GP was still an undisturbed Eden. We drank freely from the stream that tumbled cool, clean, and fresh from Gunung Palung, the park's namesake mountain canyon. The juxtaposition of abrupt mountains with lowland rainforests and peat swamps made the area exceptionally rich in wildlife, especially Borneo's endemic proboscis monkeys, gibbons, and orangutans.

HEADING UPSTREAM IN our overloaded sampan, the kids sang songs, much to the delight of the boatmen, who paddled tirelessly with short digging strokes, propelling us against the current. Over lunch, Peggy and Roman complained that it didn't feel like Borneo. But as the forest canopy closed in over the stream for good, we passed a trio of gibbons in full view, only twenty feet up in a tangle of trees and lianas.

The long-armed, short-bodied little apes were so close we didn't need binoculars to watch their nimble antics. Arboreal gymnasts, they swung rapidly through the low canopy, hanging loosely by their oversized fingers, legs dangling like those of hyperkinetic kids on a schoolyard jungle gym. Roman and Jazz were convinced they did somersaults.

Hours later, the creek narrowed to four or five feet. "Does this feel like Borneo now?" I asked.

"Yes!" Peggy and Roman answered as we passed through a dark forest where tree roots dragged in the water like dirty mopheads and long ferns twisted down from overhanging limbs. Sometimes fallen trees crossed low to the water, forcing us to get out and carry cargo while the boatmen slid the boat beneath sweepers. "These trees are so big and old, they look wise," observed young Roman as he clambered through a forest that had never seen an ax or a saw.

After eight hours, we reached Cabang Panti. Jumping out of the boat, Roman immediately found a shiny black millipede. A foot long, it looked like it had crawled out of the Carboniferous Age, three hundred million years ago.

"Dad, is it safe to pick up?"

I had asked Tim the same question the year before, so gave Roman Tim's answer: "Yes, it is. He's harmless."

As the enormous bug crawled over Roman's arms, its hundreds of legs swarmed in a miraculous wave that tickled and delighted him. Squinting closely, he said, "Look, you can see two pairs of legs on each segment!"

The camp compound was fifteen minutes away from the dock. No trees had been removed to build it, so the humidity was near 100 percent, twenty-four hours a day. The constant warmth, wetness, and shade left mold on anything not aired in the sun at least a few minutes each day or sealed up against the moisture. Immediately after waking, we put our bedding into dry bags—the kind used on whitewater river trips—to keep our sheets and clothes dry, comfortable for the next sleep.

The first night, Peggy recoiled at our accommodation. We arrived at a creekside hut just before dark to find mildewed mattresses and rotten floors. The conditions distressed her and Jazzy, but the inch-long giant ants, palm-sized hunting spiders, and other bugs that had taken over the squalid shelter excited Roman. We moved to a cleaner, drier hut the next day and Peggy soon forgot the previous night's discomforts.

On our first day hike in GP, Roman spotted an orangutan mother and baby feeding on figs high in a tree. We passed binoculars around for an hour, until she "kiss-called" at us in warning and we backed away and left. Returning to camp, we watched a yard-long giant squirrel as it bounded through a tree crown. Farther on, a hunchbacked lesser mouse deer darted across the path. The tiny deer's size, stature, and movements brought to mind a cottontail rabbit. We looked at its tracks in the trail's mud: cloven hoofprints the size of my thumbnail.

Later, Roman climbed inside a one-hundred-foot-tall strangler fig. The host tree had died and rotted away long ago to leave the fig as a natural, cylindrical ladder. Eight feet up the fig ladder, Roman found a tree frog. As he reached to catch the reddish-backed frog, it jumped, then glided away with toes spread wide, revealing oversized webbed feet as wings. It even circled back to land below him.

"*Whoa!*" he chuckled in surprise. "Dad! Was that a *flying frog?*"

"Yep! Good find, Rome! I've never seen one of those before!"

On one trip to a nearby waterfall, Roman found another unique frog called a rock skipper. A Bornean specialty, this frog clings to slimy overhangs after skipping across moving water like a flat stone thrown across a quiet pond. Somehow the boy caught one. With excitement in his eyes, he showed me the rock skipper's emerald green skin and azure blue toe pads. Afterward Roman scrambled down to a lower plunge pool and I followed.

Above a short waterfall, he moved deliberately across slick rocks. I knew his exhilaration, but also feared for his safety. A fall could break a limb or lead to scrapes, cuts, and a nasty tropical infection. I wanted to call out and warn him, "Be careful!" Instead I praised him for his good rock-climbing moves. Every parent knows this vacillation between apprehension and pride as a child reaches for independence.

Six-year-old Jazzy showed herself a natural athlete with common sense about risk. A joy to behold, this tiny towheaded girl would spring gracefully from boulder to boulder. If we offered help, she'd say, "I don't really need it, but just in case," and put her little hand in ours for slimy crossings above steep, rooted drops.

The hazards of nature—bears in the woods, tree fall in jungles, avalanches in snow country, rapids in whitewater—worry all parents who share the outdoors with their children. We were no exception. On one hike, we witnessed a huge tree limb fall from high above and strike the ground with a crash. The sight, sound, and damage were terrifying. We inspected the one-foot-diameter branch covered in orchids and ferns. It seemed best to wait out future downpours next to big buttressed tree trunks, the way we'd each crouch beneath a doorjamb in an earthquake.

Rainfall dictated our routine. Back in camp following our mornings of sweaty exploration, we would change into dry clothes. If it rained all afternoon, we would read while the kids wrote in their journals and sharpened their arithmetic playing Yahtzee. If it was sunny, we would spend the hottest part of the day at the creek. The kids dug holding ponds in the warm sand to more closely observe the fish they called needle-nose—caught on the surface—and "toe-nibblers"—caught on the bottom—in small hand nets.

"Daddy!" Jazz shrieked in joy, "come see the fish! The water's not that deep, only up to here!" Roman snorkeled around his "obstacle course" of sunken logs and sandbars. Beneath the backdrop of wild diversity, Peggy and I watched our kids at play in sunny, cool water. It felt like paradise.

At night, we holed up in our bug nets. While we were safe from malaria and dengue in the wilderness, the diversity of biting bugs equaled that of every other kind of creature and plant. Most nights we enjoyed meeting our "dinner guests": strange, wonderful, and often *giant* bugs that would fly at night into the dinner hut, attracted by its single electric light. Roman found a moth that looked like a scorpion when threatened, recording in his journal:

> I saw a moth during a huge rain storm. When I bothered him he would open his wings and lift up his abdomen, pretending it was a stinger. Cones on his head would bulge out. Fur on his legs would stick out. He was cool!

Once, I brought a glow-in-the-dark bracket fungus to Peggy and the kids to entice them into the night. We turned off our headlamps, closed our eyes, and let our pupils expand. Eyes ready, we opened them to see phosphorescent fungi glowing green in the dark. I recorded Roman's poetic description in my journal: "They look like puddles of water reflecting the night sky, except you can pick the puddle up, then turn on your light and see you're holding a rotting leaf with a little mushroom growing on it."

Cabang Panti's main building served as the pantry, kitchen, dining hall, and gathering place. A half-dozen shelves held a moderate-sized library of reference books, field guides, and Xeroxed scientific papers in binders. As in all the huts, the library was open to the humid forest air, without air-conditioning, walls, or even screening. Book pages felt damp, soft, and moldy. At night a bewildering variety of colorful cockroaches swarmed across book bindings. Some of the thicker volumes had been tunneled by termites. I pored over the mildewed texts undeterred and scribed notes in my journal to share what I learned with Peggy and the kids. It was exciting and rewarding to see such a wonderful and novel place firsthand while learning from books and articles on site.

Each day we'd walk the network of trails that crisscrossed the research area to explore the peat swamps and granite creeks. One day, we climbed to the top of Batu Tinggi: GP's summit of giant boulders covered in bright green sphagnum, serpentine nepenthes, and violet-colored flowers. The cloud forest, dripping in soggy moss and lichen, was strangely silent of birdsong and surprisingly chilly. Unfortunately, it was still full of the ubiquitous leeches eager to suck our blood.

Past Batu Tinggi, I went on to recover a compass left by a GP researcher. Peggy and the kids descended without me. When I caught up with them in a pounding rain, Roman was leading. We were excited to be reunited, even if separated for only an hour or so. "Roman's been doing a great job. He's so brave, breaking all the spider webs for me and keeping a good pace. Sometimes the trail's been pretty faint, but he's kept us on track," Peggy reported.

Roman, then eight years old, continued to lead for another hour in the rain. He only occasionally lost the trail when a fallen tree crossed it. I asked him what he liked best, what he thought was neatest about the jungle.

"The neatest thing? The neatest thing is *everything*!" Roman expressed the strong, innate interest in nature that nearly all young children seem to have. "I like how the jungle is never quiet. There's always some living thing making noise."

If Puerto Rico had initiated young Roman's fascination with the tropics, then the four trips that he would make to Borneo as a child, teen, and young man cemented that fascination in place. A dozen other trips to tropical and subtropical Australia, Costa Rica, Mexico, Hawaii, and Bhutan would send him eventually—perhaps inevitably—to Central America for his greatest adventure of all.

CHAPTER 11

Jungles and Ice

Roman and Jazz, Harding Icefield, 2001.

Most parents want to raise independent, capable offspring who still want to spend time with their family. The real test of parenting comes during adolescence, when offspring act like two-year-old toddlers in adult-sized bodies. They turn secretive, exploring nonfamily relationships that run deeper than playground friendships. Roman was typical in this regard, but fortunately he still found time for me. By his teens, our trips together had established attitudes, morals, and skills that shaped him into a useful research assistant and a competent adventure partner.

As a freshman in high school, he helped me during two months at Danum Valley Field Center in Borneo. Between three feet and two hundred feet above the ground, I dangled from ropes while handling a twenty-pound chemical "fogger" that knocked thousands of insects into collecting trays. Meanwhile, Roman learned to identify these insects from a Cambridge University graduate student named Ed Turner. Together in Danum's air-conditioned lab, with Radiohead playing on speakers plugged into an iPod, they peered into microscopes and separated Ed's bug samples into groups like Coleoptera, Diptera, Hymenoptera, and others.

Back in Anchorage, Roman assisted me with my 14,000 bugs as he had helped Ed with his. We set up dissecting microscopes across from each other on the dining room table. While hunched over scopes and our data sheets one Saturday afternoon, he said, "Looking at all these bugs and seeing all this diversity is like being back in the rainforest. Check out this praying mantis ant-mimic. It looks just like an ant!"

A few years later when the research was published, Roman would find his name in the acknowledgments of a *Biotropica* paper entitled "Arthropod abundance, canopy structure, and microclimate in a Bornean lowland tropical rain forest" by Ed Turner, two other Cambridge colleagues, and me. Roman and Jazz would be acknowledged in "Spatial distribution and abundance of red snow algae on the Harding Icefield, Alaska, derived from a satellite image" in *Geophysical Research Letters*.

The senior scientist on the snow algae paper was a Japanese scientist named Shiro Kohshima, who had studied everything from orangutans in the jungle to microbes on glaciers. He led a group of Japanese researchers up to the Harding Icefield, a 700-square-mile dome of ice in the Kenai Mountains, to study the single-celled algae that color its vast summer snowfields red. Besides red-snow algae, the scientists studied an inch-long annelid called an ice worm that feeds on the algae. The Japanese invited me along as a scientific collaborator.

My role was to collect snow algae samples and count ice worms across the Harding Icefield. I brought both kids, Roman fourteen and Jazz twelve, to ski with me. The three of us would map the red snow and count glacier ice worms on a seventy-five-mile loop during a week in August 2001.

Skiing over the Harding feels like time travel back to the Pleistocene, with ice and snow as far as the eye can see. At its center, the icefield encircles mountains known as *nunataks*, a Canadian Inuit word for "land surrounded by ice." We dragged a sled full of equipment most of each day, then set up camp early, and tossed a ring-like Frisbee for fun until the ice worms came out at dusk. Then the Frisbee ring became a sampling frame for counting. In the cool of the evening, Jazz and Roman snuggled into their toasty sleeping bags inside the tent where Jazz jotted down the counts of ice worms I called out to her. We learned that the biggest populations of ice worms with a variety of sizes lived on the red-algae fields down low; higher up on the summit dome there were no worms (and no algae); in between, we found only long single worms that seemed to be moving as if on a mission. As usual, our studies generated more questions than we answered.

For the first two nights on the Harding, we camped where three glaciers spilled off in three directions; the Japanese camped an hour's ski away on its edge. South of us, the Harding Icefield opened up in earnest: flat, featureless, blindingly white, with only distant nunataks as landmarks. The icefield is sometimes blasted by powerful, storm-driven cyclones that spin off the Gulf of Alaska at over a hundred miles an hour. These storms usually come at night, following a day of rain.

It had rained all day. We were holed up in our big dome tent where we played rummy and Yahtzee, drank coco and killed time. Roman teased Jazz. The siblings have always been close, but because she'd beaten him in cards, he used his sharp tongue to even the score.

The Japanese scientists came by to drill for ice worms and establish where in the snowpack the nocturnal annelids spent the day. While the Japanese drilled through snow and into the ice below, I stood in the wind and rain outside the tent and monitored the kids' giggles and barbs.

Kohshima pulled up a three-foot core of solid, blue glacier ice. At the bottom was a living ice worm. How it got there, we had not a clue. Perhaps it followed hairline cracks, or used a unique gland to somehow melt its way down. We scratched our heads, noting the mystery in our yellow waterproof field books. After the Japanese left for camp, the winds picked up. Strong gusts interrupted the kids' card games and the tent banged and flapped during dinner and hot drinks. We laughed at first. But as the storm bullied the tent and darkness fell, the mood changed.

"Dad," Jazz asked, "should we be worried about this?"

"No," I lied, hoping to hide my fear.

"What are we going to do?" she pushed.

I thought about that. If the tent blew apart, we would be vulnerable to strong wind and freezing rain and unable to see where to go. There was no boulder to tuck behind, no soft snow to dig into for a cave, no place to hide. Not until morning would we be able to ski to a shelter hut located five miles away at the edge of the icefield, beyond a mile-long maze of crevasses.

Eventually, bigger gusts flattened the brand-new dome tent, but, as with the faded old one on Umnak, its shape rebounded each time. The kids moved to my end, all three of us in a four-person, group-sized overbag that trapped heat escaping from our individual sleeping bags. If the tent failed, we could hunker down inside the overbag and cling to each other for warmth until morning. We shoved our cookpot and stove, lighters, compass, map and food, rain shells, extra clothes, and Nalgene bottles full of warm water into our big overbag. We wanted to be ready in case the tent should blow apart and the wind scatter its contents.

We sang songs and told jokes until the wind's roar silenced us with wet tent fabric pushed flat to our faces. We retreated deep into the muggy blackness of the overbag, where Jazz asked, "How long is this going to last, Dad?"

"I don't know, Jazz, but it should be better tomorrow," I said hopefully.

Clutching tight to each other, we fell asleep eventually and woke to a calm, clear morning. The Japanese crew came by to visit. We laughed and shook our heads, sharing stories about the storm's ferocity. They had lost a tent and the five of them had crowded into one shelter, up all night, their backs against the windward wall to keep their tent erect.

The Japanese team would stay at their camp on the edge of the icefield, repair their broken tent poles, and continue their research measuring the light reflected from snow with varying densities of snow algae that actually melt snow to survive. The kids and I would push onward, deeper into the icefield, to sample snow for algae and count ice worms. "Dad," Jazz questioned as we packed up to head out, "are we going to have any more storms like that? It was scary."

"Not like that," I reassured her. "Usually good weather follows bad. We'll be okay."

A few days later, we skied among nunataks as we crossed the rounded dome at the icefield's high point. A fog rolled in off the Gulf. In those days before low-cost GPS, we relied on compass navigation and maps. Roman held the compass like he'd learned on Umnak and kept me on course as I led the way with his direction.

We came to a crevasse field creased by cracks: some big enough to swallow a skier whole in their gaping maw. "What do I do with all this stuff again?" asked Jazz as she held out a handful of carabiners, pulleys, and ascenders that hung on her gear sling like clunky costume jewelry. The cracks were open and easy to see, and the slope flat enough that we could shuffle past. Still the potential hazard was clear. The Japanese were miles away. Both kids had ascended ropes hung from tropical trees in Borneo and Costa Rica, as well as from backyard spruce, but Jazzy wanted a refresher.

"What do Roman and me do if you fall in?" she asked.

Good question, I thought. "It's pretty safe here. It'll be hard to fall in. You'd almost have to jump into a crevasse. But if I fall in, you and Roman need to anchor the rope with your skis and throw Roman's end of the rope down to me."

We moved nervously through the mile or so of icy cracks and fissures, relieved when we reached the other side of the crevasse field, where we camped out of danger. "I don't like those crevasses, Dad. They look scary and deep."

"I know. There won't be any more. We're done with them."

I had hoped this experience would get the kids interested in more glacier travel, but it had the opposite effect. Neither would ever want to go skiing in summer again. "Nah," Roman answered the next time I asked him out on an ice-worming ski trip. "Why waste summer on the snow when we have all winter to ski?"

It was hard to argue with that.

LIKE MANY OF us, Roman shifted his interests when puberty arrived and he became more interested in adventure than natural history. Between his junior and senior year of high school, he suggested we compete together in the 2004 Wilderness Classic.

Roman's initiative to enter the race was a natural outcome of family hiking trips and the fact that Peggy and I had participated together three times ourselves. At sixteen, he figured he could meet any outdoor challenge his mother could.

At fourteen, Roman had helped me field-test an adventure race course in the Alaska Range. As he preran the course with Peggy, me, and a handful of others, Roman discovered his endurance and tolerance for discomfort on a new scale. He also discovered the excitement of swift-water packrafting. The Classic would test his boating skills, as well as his tolerance for discomfort and his endurance. I also knew from a dozen Classics that the grueling event's rugged courses destroyed participants' feet. For Roman's first race, I proposed that we use mountain bikes and packrafts to avoid "feet-beat," making the experience as positive as possible for him.

Fortunately, Roman had commuted by mountain bike, winter and summer, five miles each way every day all through middle and high school. To train specifically for the Classic, we pedaled and pushed our bikes up nearby mountains and paddled down rivers and streams in our packrafts with bikes strapped to their bows—"bikerafting." We also made a ten-day trip to the Brooks Range, where we packrafted three rivers that we linked with overland treks. Come race time, we were ready.

Thirty-five of us started the race at Eureka Roadhouse, a hundred miles from Anchorage. The finish line waited 150 miles away in Talkeetna. By the end of the first day the sky threatened rain and both of us found ourselves exhausted and butt-sore. For well over fifteen hours, we had pedaled, pushed, and carried our mountain bikes across fifty miles of Alaskan backcountry and wilderness.

We set down our bikes on a tundra shelf, high in the Talkeetna Mountains, and pulled on our puffy Patagonia pullovers.

After arranging our sleeping pads next to each other, we pressed ourselves together to share body heat, pushed our feet into our empty backpacks, and pulled our deflated packrafts over us like blankets against the rain. To save weight, we carried no sleeping bags, bivy sacks, tent, or even tarp. Before we settled in to biv- ouac a few hours, Roman reached into his food bag. "Here you go, old man," he said, grinning as he tossed me a Cadbury bar. "I didn't eat my full ration today and figured you'd need this to stay warm tonight."

"Thanks, son. That's very generous of you," I replied, smiling back. "I'll just stash it for later, in case you want it back."

After cycling, pushing, and sometimes carrying our bikes for three days, pausing only long enough to shove food in our mouths or nap a few hours, we prepared to float the Talkeetna River for the final stretch of the race. We had followed well-used grizzly bear trails to portage a canyon full of burly Class IV rap- ids. As we inflated our rafts and assessed our progress, Roman asked, "Do you think we'll sleep tonight, or just paddle straight through?"

"Up to you. It's about twenty-five miles to Talkeetna. How do you feel?" We had slept maybe eight hours of the last seventy- two or so. He looked strong, although near three in the morn- ing the previous "night" he'd dragged a bit while shoving his bike through the thick alder brush.

Roman stood up from his boat. Scratching his head with both hands, he thought for a minute, then said, "I feel pretty good." He gave me the punchy grin of a sleep-deprived adventure racer, his shoulders broad, his back straight. "I say we go for it and get this thing done. It'll be great to get off our feet and into our rafts. Here, Dad, let me help you get that bike on your boat."

Ever since Dick Griffith had pulled out his "secret weapon" two decades before, packrafting had been a staple of the Classic.

The Talkeetna River on this course presented the biggest white-water challenge of any Classic to that point. Fortunately, Roman had moved to the forefront of whitewater packrafting in the previous couple of years.

At sixteen, during his first trip down a local Class IV canyon on Ship Creek—a run that most packrafters found terrifying in the early 2000s—he declared, "This is the most fun I've ever had!" He reveled in the amusement-park excitement of dropping off Ship's back-to-back four- to six-foot waterfalls. My friend Brad Meiklejohn, one of those early white-knuckled packrafters, first met Roman there. Brad told me he had been blown away by how calmly Roman handled its whitewater. Photos of Roman paddling the creek pepper my book on the sport.

Over the next decade, Roman and I took our Ship Creek skills to the Appalachians, Brooks Range, Mexico, Tasmania, Bhutan, even the Grand Canyon of the Colorado, where he and I were the first paddlers ever permitted to packraft its length.

While we had passed the Talkeetna's biggest canyon, there were still eddy lines, hydraulics, and riffles to negotiate while top-heavy with bicycles strapped to our bows. Below one canyon wall, I watched a whirlpool grab Roman's stern. He looked startled, but in control. Leaning forward and digging hard with his paddle blades, he pulled himself free, then flashed his teeth at me in a big smile. "That one almost got me!"

After three nights, we arrived in Talkeetna in sixth place for the 2004 Alaska Mountain Wilderness Classic without a single blister. Roman's finish remains the top placement by a seventeen-year-old in the history of the race. A decade after Umnak, he wasn't only carrying his own weight and keeping up: he was wilderness racing.

CHAPTER 12

Dungeons and Dragons

Hulahula River, Brooks Range, 2004.

Roman was more than just my adventure partner and research assistant. He listened to what his mom and I said, but challenged us, too, unafraid to speak his mind. "Dad, you're pretty smart, but Mom"—he grinned—"well, Mom's wise."

Even as a kid he shared Peggy's circumspection. He shared her hairline, too: a widow's peak. He kept his straight hair short, buzzing it himself. Sure, he had a mohawk when he was eight, but by high school he had discovered that girls went for his clean-cut, Harry Potter looks with his wire-rimmed glasses and high cheekbones like his mother's. He not only resembled a young wizard, he was also smart enough to be authentic, *sans* tattoos or piercings.

While other kids watched television, Roman read books—we didn't have a TV. He read fantasy, entomology, the dictionary, even crappy books where he'd skip every second page. At nine in Borneo, he read Tolkien's *The Hobbit* in a single day, then the Lord of the Rings ~~trilogy~~* the next week, a binge that left a big gap in his journal. Later, he and his friends passed around Frank Herbert's Dune series, H. P. Lovecraft, Stephen King, Mark Twain. He read so much I wondered if it was why he needed glasses.

In high school he read science, history, economics, and texts on the world of Dungeons and Dragons, a role-playing game based on imaginative narratives and magical scoring. For years, he spent each Friday evening at a friend's immersed in DnD. Roman was a renowned dungeon master, the game-play creator, storyteller, and guide. All Peggy and I knew about it was that he left home excited to cook and share meals with a group of all ages and backgrounds. Roman cheerfully joined us for natural history and packrafting trips, but knowing that he'd developed his own identity comforted us.

Roman belonged to a creative, gregarious circle of friends who met in grade school and stayed close. At the group's center was Roman's best friend, Vincent Brady. Charismatic, athletic, artistic, Vince painted and drew, played music, and wrote poetry. The two met in kindergarten when Roman found Vince belly-down on a lawn pushing dandelions into his face, dusting his nose, lips, and cheeks with pollen. Roman asked what he was doing. Vince replied he was a bumblebee, pollinating flowers. In that moment, a beautiful lifelong friendship bloomed like the yellow weeds around them.

*Lord of the Rings is not a "trilogy." "It is a single novel... published for convenience in three volumes."

Starting in middle school, Roman threw solstice parties for this sometimes rowdy crowd, who came over, stoked a backyard bonfire, barbecued meat, and stayed up all night in the endless light of summer. A young woman from Vince's circle once wrote that Roman was known for his story-telling of surreal adventures, his sharp wit and humor, for contributing to late-night conversations, drawing pictures, wrestling and laughing, waxing abstractions into the wee hours, and joking with everyone in cuddle puddles.

Roman could handle himself on his own, too. When he was sixteen his grandpa paid for a month of Spanish language classes in Mexico's artsy San Miguel de Allende. Payment for the course included a driver who would meet him on his arrival in Mexico City. But Roman's flight from L.A. was late and the driver left without him.

Roman called home. It was late in the evening.

"Dad, my flight out of L.A. was delayed, so I missed my ride in Mexico City. I bought this phone card but it's only got five minutes. What should I do?" We had to solve the problem under the time constraint of his card.

"Hang up and call the school. Ask them what their advice is. Then call me back and tell me what they say."

He hung up and I waited. A few minutes later the phone rang again. "Dad, they said that there's a hotel in the airport. I wouldn't have to leave and they'll send the car in the morning. Or I can take a bus. There's one more tonight and it leaves in less than an hour. It goes to another town where I get a second bus to another town and then I get a taxi. They said the whole trip is three or four hours."

"Where do you get the bus?"

"Outside the airport. If I leave the airport, I can't get back in for the hotel."

"Do you have all the directions written down?"

"Yes."

"What do you want to do?"

"I want to try and catch the bus."

"Okay, Son. Good luck. Call the school and tell them what you're doing. And then call me back when you get there."

Of course, he made it. And so began his first Mexican adventure. He grew up on that trip and came back a young man full of rich, humorous, self-effacing stories. We were still close; he still wanted to do things, but there was a noticeable shift toward independence.

SOON AFTER, AS a junior in high school, Roman was selected to participate in the school district's gifted mentorship program. Genetics had interested him since middle school when he had read the *Cartoon Guide to Genetics* and pronounced, "I'm just a genetic enhancement of you, Dad." I introduced Roman to a colleague, the lead scientist at the U.S. Geological Survey's Molecular Ecology Lab. She would mentor him his junior year, beginning a ten-year relationship with the lab. Running polymerase chain reactions, sequencing genes, and reading gels helped Roman pay for college and graduate school.

Working in the lab, he often listened to NPR. I asked him what his favorite show was. "*Marketplace,*" he said. "I like hearing about how the economy works."

He entered Virginia's College of William and Mary in 2005 as an economics major. He met his good friend Brad in an econ class their first year. It seemed fitting that Roman, the son of an ecologist, would be drawn to the mathematics of human ecology, but by the end of freshman year he said he didn't have a feel for economics like he did for biology; he changed majors. He also met his first serious girlfriend his freshman year in an art history class.

Roman had kept his high school relationships with girls to himself, but the young woman he met at William and Mary was different. Following their junior year, she came to Alaska for the month of June. He shared with her the adventures he had grown up associating with family—backpacking in Denali, sea kayaking in Prince William Sound, packrafting Eagle River—a clear signal that he was serious about this girl. We had never seen him smile and joke as much with anyone. I was happy for him and thought he'd found his soul mate.

After graduating in June 2009, he returned to Alaska with a biology degree and this college sweetheart. The two moved into an Anchorage apartment right after graduation. All of his friends and ours delighted in his girlfriend. But the following spring—during the Alaskan season known as "break-up," when the ice and snow melt and the rivers run free—she ended the relationship. Soon after, his friend Vincent Brady died from an extremely progressive form of cancer. The dual loss left Roman heartbroken and crushed.

One day, he came by the house. He stood at the door, the pain on his face like he'd been physically kicked. I gave him a hug and asked how he was doing, if he was okay.

"How do you think I feel, Dad? My girlfriend left me and my best friend died. I feel stormy and difficult, mean and sad." His world painted black, I didn't know how to recolor it, beyond suggesting obliquely that someday, somehow, he might reconnect with her.

After the split, he moved in with a roommate across town. I came by to drop off his things from storage at his new apartment. Three upside-down bottles of hard alcohol stood on tap in the kitchen: vodka, bourbon, and tequila ready for an easy drink. He said the liquor was his roommate's. I had my doubts: in his room there was a big dent in the sheetrock.

"What happened here, Roman?" I asked.

"I hit the wall with my fist," he answered matter-of-factly.

"How come?"

"I dunno. I was mad?"

No doubt he was mad. Mad with his ex, mad with the world for taking his best friend, and mad with himself, helpless and hurt with no apparent way out.

That summer, Roman collected specimens in arctic Alaska as one of a trio of scientists studying the effects of climate change on small mammals. His notebook recorded solo hikes and hunts each day. Neat, legible details filled page after page: the animals he saw and caught, the coming of autumn, the game trails he followed, and the bears, caribou, and wolves who made them. "Anticipate getting more voles later in the season as food begins to run out," he hypothesized, an idea supported by counts of small mammals that he recorded next to his hand-drawn maps.

Roman didn't mention his feelings of loss for Vince or his girlfriend. His notebook's entries focused outward, ignoring the grief and turmoil inside. He described hunger and the satisfaction of good meals; sore feet and the discovery of good walking. Most outdoor adventurers turn to the wilderness after emotional loss; others turn away. For Roman it distracted from the pain of losing the girl he had most loved in his life and the friend he most respected and admired. Time would mend his broken heart, while comfort, love, and companionship could speed its healing.

WE REGROUPED AS a family during Jazz's Christmas break from her third year at Lewis and Clark College in Portland and returned to Borneo in December 2010. Both Roman and Jazz recalled vivid childhood memories there. We went to new places, including an island resort in the Celebes Sea. Both kids were tickled to sleep with air-conditioning and eat rich, gourmet food. We participated in organized activities—even tried karaoke. At least Peggy could sing; my croaking put the kids into stitches. The staff held a competition for guests: Roman used his hands to put makeup on Jazz from behind her back, unable to see what he was doing. By the end her face looked like a Jackson Pollock painting. The two of them nearly collapsed in laughter. It was good to see Roman happy and having a good time.

Back from Borneo, Roman entered the master's program at Alaska Pacific University and threw himself into a sophisticated thesis project building an evolutionary tree for the thumb-sized isopod *Saduria entomon*. He continued to struggle with the loss of his girlfriend, writing in his journal while doing fieldwork in Alaska's arctic:

Of course everything hurts . . . I don't want any of these feelings. I don't like being sad or feeling crushed. I'm obsessed and angry and feel so vulnerable. No reason to write down my feelings. They're boring and I don't want to feel them twice. Not drinking is hard. This trip will test me, I think. I would like to see it through. See something through . . . I keep telling myself I have a high threshold for discomfort. Not sure what that means though. Need to do more outside. This trip is awesome, but I need to be moving more. Not hard enough? Need a hard solo trip, to remind me I'm weak but alive.

He was looking for something other than alcohol to resolve his lasting feelings of loss. Returning from the field, Roman worked on two scientific papers published in 2012. The article in *Molecular Phylogenetics and Evolution,* "Historical biogeography of the North American glacier ice worm, *Mesenchytraeus solifugus,*" was a particularly daunting manuscript for a young scientist who'd not yet completed his master's degree. The other was a technical note on the genetics of snowy owls. Roughgarden was right after all: at twenty-five, Roman was the lead author on both published papers, a bona fide biologist.

Roman's friends, colleagues, and fellow travelers later reached out to us. Their emails and conversations helped us to see what kind of man he had become, beyond our family's perspective. As one of his friends wrote, After Vince was gone, the emptiness settled in. Roman seemed to maintain the vibrancy of that atmosphere we had all shared with each other, his exuberance, and I was so thankful for that. Another elaborated on a gift Roman brought her from Bhutan:

> It was after Vince's death that Roman opened his heart to me in a way I had not experienced previously. He was cooking dinner and he had a gift for me. When I arrived, he pulled me aside, and he placed a set of prayer flags in my hands. He looked me straight in the eyes, and said, "I owe you an apology and have for many years."

While Roman had a tender side—nursing me to health once in a Bornean hotel while I recovered from a tropical fever—he also had a cynical one. As a graduate student at APU, he'd made friends with a group of students who'd go on to medical school and other professions. In a touching email to Peggy and me, a friend named Don Haering described him:

> . . . an unusually intelligent and interesting character. I loved interacting with him, as nearly everything he said was thought-provoking in some way. He was the type of person who made me think carefully before I spoke, as I knew that he would probably have a well-informed question

or response. Not only was he usually the smartest person in the room, I think he made everyone around him a little bit better too. In class, or any group discussion, he had a way of listening quietly and letting the conversation play out, before delivering a comment or response that was always on point, and which often completely reframed the dialog. It was a little skill of his that I came to anticipate, enjoy, and which I still attempt to emulate. Whenever I spoke with Roman personally, I always had the sense that he was amused by the world around him, like there was humor in every situation, even the mundane. In that way, as well as his obvious sense of general curiosity towards the world, I felt that he was a kindred spirit. I feel fortunate that we crossed paths.

Don's moving character sketch confirmed how Peggy and I had hoped Roman would turn out: informed, influential, equipped with a sense of humor. It satisfied me that even the smart kids saw him as a role model.

Don was also going pretty easy on Roman: "reframing the dialog" often meant challenging disagreement. Roman sometimes found me a bit too sentimentally liberal, for instance. But for all of us in the Dial household, the actual differences in our opinions are less of a problem than our similarities in the way we disagree: disagreements often escalate, but subside just as quickly. Nobody holds a grudge for long.

In 2012, when Roman and I walked out after searching Bhutan's Himalaya for the Tibetan ice worm, we followed a trail that led to a remote village called Laya, perched in a picturesque valley pushed hard against the Tibetan border. At the time, Laya's two-story stone and wood homes were off the grid and days from the road system in a wilderness where people lived.

Leaving Laya, we encountered laborers and horses ferrying power poles and spools of cable. I complained that the arrival of electricity would kill the village's charm. Roman accused me of projecting my sentimentality onto people who deserved the con-venience of electrical power. I responded it would dilute their culture. He retorted it was up to them, not us. For miles, we each stammered in frustration as emotion eclipsed logic, each of us clinging stubbornly to our side of the argument.

All fathers readily see their foibles reflected in their sons, and there, plain as day, were mine.

CHAPTER 13

Big Banana

Twenty-footer, Rio Alseseca, Veracruz, Mexico, January 2014.

Roman dated Katelyn, another APU student, in 2012. Working with her on a project to estimate small mammal abundance near Anchorage, Roman taught her techniques he had learned during his previous field seasons up north. A year later, when his computer simulations of isopod evolution didn't converge on a solution, he decided he needed a break.

He settled on heading east to visit college friends, followed by a bicycle tour through Kentucky's bourbon country, then a long-term sojourn through Latin America. In October 2013, with his student loans paid off, $15,000 in savings, plans to spend Christmas with Brad, and enough Spanish to travel to South America, he told Katelyn he was breaking off their relationship. They remained close, though, and she joined him in Mexico for some packrafting and Maya ruin exploration in early January 2014. Shortly after she headed home, I met up with Roman in Veracruz, a state in eastern Mexico, to packraft with him and a handful of our Alaskan friends. We both looked forward to doing the kind of things we'd been doing together for decades.

Roman greeted me at the Veracruz airport. He was a month shy of twenty-seven. He had gone a few weeks without shaving and his new beard accentuated the lean angle of his jaw. He certainly had his mother's good looks and the scruffiness didn't hide them.

Roman was up for some whitewater adventure on what's been called the "best bedrock" in North America. Kayakers come from around the world for Veracruz's steep, polished gorges, vertical waterfalls, and tumbling cascades. We were eager to go paddling, but first we had to get something to eat.

I rented a car and we drove off into the coastal city of Veracruz looking for good Mexican food, maybe some *carne asada* tacos, or "street meat" as he liked to call it. He was excited. We hadn't seen each other for months and had a lot of catching up to do. He told me what he'd been doing, where he'd been, about his travels with Katelyn across the Yucatán. His words poured out. The son of a noisy father, he tended to be quiet, so when he spoke, I wanted to hear all that he had to say. Besides, we'd soon pick up two more boating buddies, including my good friend from Alaska, Brad Meiklejohn, who's my age. When they showed up, Roman would listen more than talk.

Our friends arrived the next afternoon and we headed inland to paddle warm whitewater. We did a day trip down a limey creek that issued full of life from a hillside spring to twist and turn through open woods and pastures where Brahma bulls laid in the shade. Packrafts have come far since Dick Griffith unrolled his pool toy in the first Wilderness Classic. Three decades on, they look more like fat little kayaks than small round life rafts and increasingly imaginative boaters paddle them down whitewater creeks and rivers normally kayaked or never run previously at all. Many experts can even "Eskimo-roll" their packrafts: if a rogue wave or turbulence flips the boat, the paddler rights it and paddles onward, all while still in the boat.

After the lime springs creek, we drove to a town called Jalcomulco, where we hoped to find a local who'd shuttle us to the put-in of the "Grand Canyon" of the Rio Antigua. Unfortunately, all of Jalcomulco's boating community were busy protesting a proposed dam that would flood the Antigua's canyons. We'd have to drive ourselves and leave the rental car at the put-in during our overnight trip downstream.

We enjoyed our paddle down the clear, moderate Class III waters set deep in lush, green gorges. We camped in the woods on the river's banks and warmed up the selection of Mexican foods that Roman had picked out for us to eat around our crackling campfire. Tenting with him was familiar and we fell into an easy routine.

A second full day of bigger water brought us back to the town of Jalcomulco, where we spent the night, then drove to retrieve our car the next day. Returning to the rental parked at the canyon put-in, I was puzzled to see the Volkswagen sagging with a door wide open. Pulling closer it was apparent that all four wheels were gone. So were the battery, the carburetor, the radio and CD player, and the few items we'd left in the trunk, including Roman's empty backpack. Roman would eventually replace his pack with one he'd buy in Mexico after the rest of us left;

until then, he borrowed Brad Meiklejohn's. As with any theft, we felt violated, frustrated, and hurt. The episode cost us a day or two but soon we were off to Tlapacoyan, the center of the Veracruz whitewater scene. Adrenaline has a way of washing away unpleasant feelings.

The highlight of our two weeks paddling packrafts on rocky streams was an exciting descent of the Rio Alseseca's "Big Banana," a steep creek rushing through the jungle.

Even though our friends from Alaska, Todd Tumolo and Gerard Ganey, had completed the committing run down the river before we arrived, I was nervous for Roman. He hadn't been on the likes of the Alseseca for more than a year, and while we had worked our way up to the Big Banana's challenging Class IV waterfalls on easier runs over the previous ten days, I was still concerned for his safety. He was, after all, my son.

Ganey, Todd, and other friends all said the Big Banana was the best whitewater run in the state of Veracruz. It sounded thrilling and relatively safe to me. We could easily walk around its biggest, most dangerous waterfalls, thirty and forty feet tall. But I wanted Roman to feel good about it and have fun, too.

Despite our collective experience and our three friends' run the week before, our descent of the Big Banana nearly ended in its first hundred yards. Ganey, an expert paddler, decided to try a short, messy cascade that poured through a jumble of boulders. The rest of us had already walked past the hazard because it didn't look very "clean." Clean rapids don't trap and potentially drown a swimmer like "chossy" ones can. We positioned ourselves below the drop with safety ropes.

Ganey paddled smoothly into the rapid's entrance, maneuvered off the lip and prepared for his landing. But instead of plopping smoothly down, he was grabbed by the rapid's rocky edge and flipped out of his boat. Almost immediately, a whirlpool sucked Ganey underwater into a sieve of boulders where the river's hydraulics brought him back to the surface, only to shove him under again in a recirculating current.

The chossy drop wouldn't let him go. He cycled around and around, fighting for air and his life. I threw him a rope but the Alseseca swallowed him again before he could find it. Fortunately, Todd's throw line followed mine and Ganey grabbed it the next time he resurfaced. Todd dragged him, exhausted, from the current. We all breathed a sigh of relief.

Watching Ganey there, splayed on a rock heaving for air, an unease informed my judgment. Although a practicing scientist and college professor, I've learned the hard way never to ignore intuition, either mine or others', especially when it involves my offspring.

How about the other rapids downstream? How safe are they? I wondered. We were able to walk around this drop, but if we would later be forced to paddle dangerous cascades like the one that grabbed Ganey, then I was ready to pack up and head right out on the dusty trail we'd followed in.

I turned to the others. "What do you think, Roman?" I asked.

He looked cool as a cucumber. But I knew my taciturn son could hold back his emotions. Roman had watched Todd pull Ganey from the water trap. "I don't know—*that* drop looks pretty hairy," he said, shaking his head slowly. "It's why we're all putting in below it, right?" Unlike me, he had never been accused of being an adrenaline junkie.

"What's it like downstream?" I asked Todd. Ganey's near miss had been on a short, five-foot drop on a big creek. Twenty-foot waterfalls, impossible to portage, waited below.

"Oh, it gets better. Much better. This is the chossiest drop on the whole run. It cleans up."

"Are you sure?" I wanted to know.

In adventurer's slang, Todd denied he was downplaying the river's challenges: "I wouldn't sandbag you, Roman."

I turned to my son to read past his composure. We had pack-rafted whitewater together for over a decade. He knew when to say no. He knew when to walk. I had made it a point never to force anything on him that he didn't want to do. He would let me know. While he was often quiet, he was never shy. He was his mother's son, as much as my own: risk-aware, vigilant, never hesitant to tell me what he thought.

"Well, Rome?"

"I say we go for it, Dad."

"Okay, then, I guess that's settled." I looked at the others. "Todd, you want to lead? Ganey can you bring up the rear and run sweep?" They nodded and we got into our colorful little boats.

"Let's go."

We slid into green water that flowed smoothly over dark rock as canyon walls rose overhead and closed in all around us. Broadleaf crowns of contorted jungle trees dangled from the rim above as we followed the twisting, crystalline creek. Sometimes we shot over rocky underwater ledges where we paddled aggressively off powerful pour-offs. Other times we performed aquatic pirouettes as we maneuvered through tumbling cascades, stabbing our paddles into the current to turn us abruptly and dodge the obstacles. We enjoyed the Alseseca's challenging rapids and welcomed walking around its dangerous ones. I felt parental pride watching Roman, back to his old whitewater form, negotiate drops with skillful strokes of his paddle.

Midway down, the Alseseca River rushed through a narrow gorge to plunge off a twenty-foot waterfall that was impossible to portage. We surrendered to the drop, whooping as we launched off its edge, falling with the water into the warm, clear pool that yanked us all from our little boats with the force of our entry.

Roman and I clambered back into our packrafts. After catching our breath and soaking in the endorphins, we laid back and looked up. The pool was set deep in an overhung alcove, a natural amphitheater. There was no way out but down. The next drop, while falling a vertical twenty feet, was nowhere near as violent: we'd slide down it, not plummet like we just had.

"That was something, huh, Roman?"

"Yea! That was crazy! There was no way I could stay in my boat," Roman said. "When I hit, the current just ripped the boat right off me and my thigh straps. It felt like someone forcibly pulling off my pants!" He laughed at the recollection, exhilarated by the thrill. "I was really nervous going off, not being able to see where we were landing. And it was a long fall! But WOW! A twenty-footer!" He shook his head with a look that said he felt vibrantly alive.

Despite my concerns at the put-in, the Big Banana turned out to be a fitting end to our two weeks together. Exhilarating but safe, the run felt like an amusement park ride, albeit with consequences, like Ship Creek only far, far bigger.

WHEN I HEADED home to Alaska, Roman came to the airport to see me off. He grabbed my black duffel full of boating gear from the rental car's trunk, threw it on his shoulders, and hauled it to the terminal. As we made our way to the check-in counter, he told me about his plans to head overland to Brazil. He would start with a trip into Mexico's Sierra Madre to see millions of butterflies roosting in tall pine trees—nearly the entire population of monarchs overwintering at the end of their migration.

There were guided tours available, but, he said, "I'm going to find the monarchs on my own."

That a boy—chip off the old block, I thought, grinning. He'd been raised on trips of independent discovery where we used our wits, knowledge, and experience to explore the natural world. It was good to see him continuing those kinds of adventures on his own.

He set my bag down at the check-in counter and spoke to the attendant in Spanish. He turned toward me and I told him, "Good luck, Roman. Have fun. Be safe and stay in touch. Mom and me will want to hear about those butterflies and everything else."

"I will."

I pulled him in for a hug. "I love you, Son."

"I love you, too, Dad." He smiled and I turned for the gate to head home, happy to have spent this time with him in wild nature and looking forward to what his next adventures would be.

PART II

El Petén

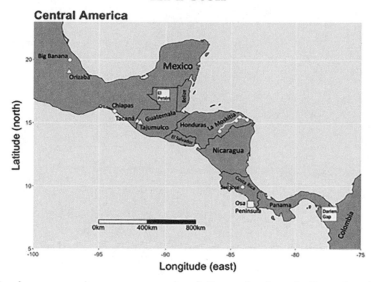

Triangles are mountains, squares are regions, bullets are locations, the diamond symbol is a beach, and the white line is the Patuca River.

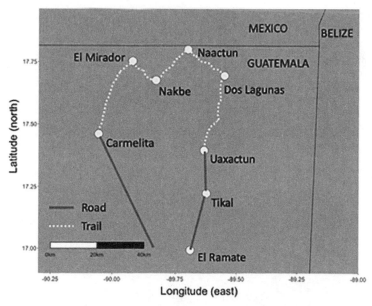

The M-shaped route. Bullets indicate place names in the text.

CHAPTER 14

Mexico

Dry forest, Guatemala, April 2014.

After we said good-bye at the Veracruz airport, Roman stayed in Mexico. He climbed its highest peak—Orizaba—the first week in February, then searched out the overwintering monarch butterflies in the Sierra Madre soon after. Peggy and I didn't hear from him for a couple of weeks, until Todd posted a YouTube video of our trip. Roman replied-all with one word: "Bitchin." He lamented sending his boat home with me.

I really miss my boat. Theres been a bunch of stuff that would have been great with a packraft, and Guatemala and Honduras are full of rivers. Costa Rica and Colombia, too. Should have sent the tent back with you and kept the raft. I met a German who was traveling around with his parapent. Hes up in Michoacan flying right now.

As Roman traveled farther south, he emailed us more frequently, apologizing for the typos and absence of apostrophes in emails written on Spanish keyboards. I was teaching full-time, writing research reports, and working on a remodel and insulation project in our attic. Hearing from him brightened my busy days as he described the places he went, the people he met, and the foods he enjoyed.

The night before Roman's twenty-seventh birthday, on the west coast of Mexico at an off-beat beach town in Chiapas, a thief stole his iPhone, buried beneath dirty laundry and toiletries in his blue Kelty tent. Because texts and international calls were too expensive, he rarely used the phone except for Internet, music, and occasional photos. He didn't notice it was missing until morning. He emailed me immediately to cancel his account before any charges were made. I cringed knowing a thief had robbed him on his birthday. Roman would not replace his phone.

Roman had bought a pack to replace the one stolen in Veracruz and returned Brad Meiklejohn's to him by mail. Roman complained that his new Mexican pack—stuffed with his tent, cookpot, Jetboil stove, cold-weather gear for climbing, and a yellow waffle-surfaced sleeping pad strapped to the back—made him a mark for hucksters. As he traveled south, he would use it for storage at hostels when he went off to climb volcanoes and canyons. On those excursions, he carried a small yellow duffel bag over his shoulder bike-messenger-style. Our friend Forrest

McCarthy had marked the bag with his name and Jackson Hole address and given it to us years before.

Roman was disappointed with camping in Mexico. It was overpopulated, polluted, and dusty. He had enjoyed clean drinkable water in wilderness all of his life. But in Mexico, he wrote, "the water needs to be treated, and everything is downstream of something you dont want on you." Between the thieves and the "cow shit everywhere," he'd had enough of Mexico.

"What's next?" I wrote back, curious and excited to keep track of his adventures. He said he planned to continue overland through Latin America, maybe as far as Brazil for the World Cup in July. There would be volcanos and jungles in Guatemala, the Blue Hole in Belize. Then, on to Honduras, cheap, but also one of the most dangerous countries in Latin America. He planned to surf Nicaragua's Pacific waves, visit Corcovado in Costa Rica, Panama, Colombia, Peru. His itinerary sounded adventurous, but I worried for him, too, in countries known for their desperate conditions and crime. He closed his email with Love you guys, thanks for teaching me important life skills. It was the kind of admission all parents yearn to hear from their offspring.

EVEN THOUGH ROMAN had studied Spanish through high school and taken the immersive month-long course in Mexico at sixteen, he struggled with rural Mexicans' thick accents. Still, his command of the language improved as he went:

My spanish is good enough now that I can order a banano liquado [a fruit drink] at a cafe, be told there are no bananas, ask if I get bananas they'll make one, leave the cafe, realize I dont know where to buy bananas, ask a nearby shop owner where I can buy bananas, get directions, buy the bananas, return to the cafe, and negotiate a lower price for my liquado.

Roman seemed surprised in his ability to navigate and re-member places and people, skills he had not noticed in himself before: it's kind of funny, 'cause I've spent so much time following you around, Dad, with your innate sense of direction . . . that it's really just been a failure to pay attention.

He described tricks he had learned. He made it a point to walk through a town to get a feel for its layout in his muscles and bones and to watch for street signs, mileage markers, and landmarks when on public transport—a challenge for passen-gers who don't need the kind of engagement that drivers do. With Latin American cities and towns mostly aligned north-south, Roman learned to triangulate his position using nearby moun-tains or tall buildings as cardinal landmarks. He found that locals willingly offered directions but mysteriously resisted using street names: I have no idea why they don't just say 22 and 6 instead of "go left, one block, then right, four more blocks, and across from that thing." I was encouraged that he was discovering *how* to travel and honored that he shared his discoveries with me.

Because Alaska's wilderness rarely has marked or maintained footpaths, Roman and I had usually followed trails made by moose, bear, and caribou. To find and follow their routes takes a sixth sense developed through experience. Just like following game trails, Roman wrote, people trails and streets have a similar intuitiveness. He found when climbing volcanos that the widest trail was usually the right one to the summit. It was good to see that all those miles he had dogged behind me in Alaska were helping him elsewhere, too.

While Roman found physically negotiating the towns and countryside simple enough, there was a darker side to navigating

Latin America, too. Almost every gringo he spoke to who had lived in Latin America was forced to leave after a year or two, due to local hostility or pervasive corruption. One woman, who had lived twelve months in Belize, told him he would be safest if he hired guides for explorations. But while guided tours were safe, even cheap, he wrote, they weren't as fun as solving his own geographic puzzles. To climb Orizaba, the tallest mountain in Mexico, he researched online for a week, then headed up alone while trailing some locals who carried ropes and ice axes. He wrote that in the Sierra Madre:

> Finding the butterflies was super fun, as I had no idea where they were. I reasoned that the butterflies were up high, so I navigated towards the highest mountain along a myriad of forest paths. Going up, I figured that the biggest path with the most horse crap would be the tourist path. And it worked perfectly.

I wasn't surprised Roman eschewed guides. After twenty-five years of travel with him, I could count on one hand the number of times we'd used them—a night walk in Australia looking for tree kangaroos; scuba diving; on a wildlife-watching tour from an eco-lodge in Borneo; in Bhutan where all foreigners must be guided.

Following Orizaba, Roman climbed more volcanoes, including Tacaná, a 13,800-foot mountain that straddles the Mexico-Guatemala border. Because alpinism had been both so addictive and dangerous for me, I had purposefully resisted introducing it to him. But climbing volcanos is, in essence, uphill hiking at high altitude without the objective hazards of falling rock, ice, or snow—or the subjective ones of falling off a cornice or into a crevasse.

A French traveler had recommended Tacaná and suggested

Roman hire a guide to do it, probably because land mines from previous conflicts were rumored to booby-trap its slopes. Roman ignored his advice. Instead he asked a hotel desk clerk about the route while near the base of the volcano in Mexico, bought water, chips, cookies, and chocolate, then caught a *colectivo* (the cheap minivan used by locals as transport all over Latin America) at dawn that took him to the end of the road. There he followed the obvious trail leading up.

As on nearly all of Central America's mountains, *campesinos* farm Tacaná's slopes, where they grow potatoes and beans and graze cattle and goats. Near villages, the trail braided out confusingly and he asked the friendly villagers where to go. Their Spanish directions led him over the border into a clean, Guatemalan village. Next, he followed goat trails marked by little cloven hoofprints. He climbed higher into the clouds, where, with limited visibility, he scrambled over boulders and wandered through tall pines filled with birdsong. Unlike rural Mexico, there was no trash, and little sign of humans. "It was nice to feel alone for a little while and the white-out clouds obscured the myriad villages below," he wrote. "The air was fresh, too, above the perennial smog that hangs over Latin America."

By the time he reached the summit cone, a thunderstorm hastened the arrival of night. His headlamp beam bounced uselessly against the mist, so he picked his way down in the fog and darkness without it, unsure where he'd arrive—Mexico or Guatemala—at bottom. With farmers in bed at sundown, there was no one around to ask for directions.

Luckily, I've followed Roman Dial around the wilderness all my life, so my instincts led me to the right place. I missed the last colectivo, but caught a crowded taxi. Driving was terrifying, since the same problem with my headlamp was 100-fold with the headlights. Three of us hung out the windows in the rain and clouds shouting "Derecha! Derecha!! Izquerda!" to keep from going over a cliff.

The next day he bought a pound of local Chiapas coffee for 50 pesos—about $2.50—loaded it in his pack and left Mexico for good. He entered Guatemala with his clothes smelling like the fresh grounds: "Amazing," he wrote. Guatemala, Roman surmised, was "a legitimate Third World country," recalling how Indonesia, rural Malaysia, and Bhutan had felt, sounded, and smelled. Stories of robbery and murder, he said in an email, gave Guatemala an edgy feel.

Roman soon set out to tackle the highest mountain between Mexico's Orizaba and Colombia's Andes: Guatemala's 13,845-foot Tajumulco. To increase the challenge—and hence the reward—he decided to skip even the Internet and guidebooks. He wouldn't ask any Frenchies, he emailed, but just ask locals, as a good way to sharpen his language skills. His plan was to navigate the rural confusion of dirt roads, trails, and farmlands with just his wits and his Spanish. It would be a different sort of hard and risky, he said. *And more like true exploration*, I thought.

CHAPTER 15

Guatemala

With friends, San Pedro La Laguna, Guatemala, March 2014.

After his almost three months of traveling alone and translating for others, Roman's conversational Spanish edged toward proficient. Knowing only that the closest town to Tajumulco was San Marcos, he befriended an old cowboy campesino who showed him where to catch the bus. In San Marcos, he asked for directions from a group of middle-aged women. With his mother's cheekbones and heart-melting smile, he had good luck with this demographic. "Plus," he said, "they had great directions."

Roman found his smile went a long way in Guatemala, especially with the Maya. The Maya reminded him of the Malays in Borneo. They were short-statured, friendly, and smiley, with no outward aggression—except the Maya seemed more willing to thieve than the Malays. Thieves or not, it was clear from his emails during his seven months of travel that he preferred Guatemala over all the other countries he visited.

Near Tajumulco, he jumped in a cab with five locals. Dropped off at a hotel, he went to the desk where three "tittering little girls" and a teenage boy checked him in. Roman tried to pry information from them about the climb, but couldn't understand their directions. Then the children's charismatic father appeared and in "tourist" Spanish gave "spectacular directions" (which I made him repeat about a dozen times).

Hungry after a long day, Roman asked where he could find food. "The bossiest little girl took me to the house next door and asked the grandmother there to make dinner." As he waited, three little boys grilled him in Spanish about everything, which, he said, was fun. Over the course of the meal, the entire extended family trickled in. An older brother had spent eight years in the States, so he and Roman traded stories, each practicing their second language on the other.

Roman easily made it up Tajulmuco's straightforward route to the top, but was disheartened to find the mountain's forests overgrazed by goats, its trails littered in trash, and the summit views obscured by cloud. He had approached and reached the summit using nothing more than his Spanish language skills in gritty, rural Guatemala. That was the fun part. As parents, Peggy and I admired Roman for this creative approach to travel in the Internet age. He went beyond securing transport, food, and lodging in a foreign land. He relied on his command of a foreign tongue and his ability to interact with strangers to find his way.

Roman was as thrifty abroad as he was at home. Equipped

with local knowledge about prices, he bargained before committing to any purchase. "Money talks, loudly, in Guatemala," he wrote, where budget travel made him less of a target. He traveled for a spell with a British woman whose parents were Scottish, observing that "She is exceptionally stingy. Which is great, because we can practice our Spanish negotiating prices."

Peggy emailed him about making friends. He described the budget travelers who wanted to do things, but who were too intimidated without a guide and too cheap to hire one. He enjoyed playing the role of trip leader for these newfound friends and described his approach in detail: First, it was important to be friendly: "Hey, let's go get food." Next, inclusive: "Hi there, you've been here a month, how do I do this, what's it cost?" Then offer a suggestion: "I hear there're free salsa lessons tonight, want to go?" or a bit of advice: "Here's how you climb X on your own, but you might not get as good a view without doing a tour package that leaves earlier because of clouds." Now, set the hook with an invitation: "Want to go to the hot-springs tomorrow? It's easy and safe." Finally, lead them there and escalate: "Hey, the hot-springs were fun, weren't they? How about climbing a volcano? It has a sacred lake in the crater surrounded by Mayan shrines. No, we won't need a guide. We can just go there and ask some people."

It sounded like Roman was far from a lone wolf on a solo adventure. One traveler he met—who wrote us later—remembered how helpful he had been to her:

> Roman was always so knowledgeable when he was showing me around or translating for me and I will always remember how cunning and strong he is and all of the valuable lessons he taught me. . . . He took me under his wing and took care of me since I was traveling alone in Guatemala, and as I have told him, I will be forever grateful for crossing paths.

BY MID-MARCH, AFTER he had spent a week or so around Guatemala's Lake Atitlan, climbing volcanos and visiting hot springs, Roman planned to head for Tikal, the ancient Maya's most iconic ruin. There he'd talk to people about El Petén, the biggest Latin American wilderness north of Panama and full of Mayan ruins. He identified his next adventure there in northern Guatemala near its borders with Mexico and Belize.

"Dad," he wrote me in mid-March 2014, "there's an undeveloped Mayan ruin 63 km into the jungle. Everything I've found says get a guide and mules because there's no freshwater and carrying water that far is impossible." At eight pounds a gallon and one gallon a day, Roman couldn't carry much. He'd be too heavy and slow. Apparently five thousand people do the trek a year, so he wasn't worried about pathfinding without maps: "If I get lost I just turn around and follow my blazes out. What do you think? Hump 12L out and see what drinking swamp water is like? No good, head back?"

I was flattered he asked my opinion and pleased to see his evaluation of the risks. While we had spent months in rainforests in Asia, Australia, and Central America, other than a week-long walk across Corcovado when he was eleven, we had generally stayed at research stations, where we slept in huts or base-camp tents and made day hikes in search of animals and plants.

Our self-propelled camping adventures, where we moved by foot, boat, bike, or ski across a hundred miles or more, had been mostly in temperate, boreal, and arctic landscapes. He could certainly boil water for drinking as we had at the bottom of the Grand Canyon and across cattle lands in Mexico and Australia. Unfortunately, fuel for his Jetboil stove had been difficult to find. But because it was the dry season, he said, fires wouldn't be "equatorially hard" to start, a nod to his experience with 100

percent humidity and daily afternoon rains at the center of the
Earth's tropical regions on the equator.

Roman admitted that safety from criminals was his biggest
concern. Like most of the world's international borders that are
largely wilderness, El Petén has its share of bandits, hostile locals
resentful of outsiders, and narcos transporting Colombian cocaine to Mexico. Even tourism could be dangerous. He met one
young traveler who had witnessed a "tourism cartel" send armed
men with guns to chase down tourists and ensure that they hired
the "right" *coopertiva* guides. Guiding was a big source of cash for
the rural economy; nearby narco traffic and access to weapons,
apparently, had encouraged its criminality.

He had asked for my advice, so I wrote back. If thousands of
people did the trek each year, he wouldn't really be alone.

> I expect that you'd be able to find fresh water—there's no way to carry
> enough. I like your walk in with two gallons. If you find none the first
> night you have enough to walk out. Ideally you find water coming out
> of limestone cenotes [sinkholes]. If it's moving and there's no trash or
> people around, it'll be pretty good. Otherwise boil some at night in
> camp, let it cool and carry it the next day. Boiling swamp water is OK. I
> am pretty sure you will find water.

My email reminded him about tinadizole, the one-pill treat-
ment for giardia that we would buy in developing countries
where giardia and dysentary are common.

I'd say go for it, I wrote in closing.

But two weeks later his plan had changed. Now he described
a new route, much longer and far more remote. He'd be gone ten
days. He lacked a good map of the area's "thin jungle trails" and
would follow them with little more than a compass and his wits.
I shared his apprehension:

I expect I'll spend a couple days out there, eat a snake, get scared, and turn around. If all else fails, I can always just walk south and hit a road. The distances here are quite small. Honestly, I'm more worried about dealing with the tourism cartel on the mule track from Carmelita to El Mirador than I am of getting lost in the woods.

Roman had designed a tent for his trek, bought material, and had a Guatemalan child sew it together. He looked forward to seeing how his design would work. I'm also pretty excited about getting to use a machete, he wrote. He promised to update me once more before he left and to leave his plans with a local ex-pat in the nearby town of El Remate. But reading and rereading the email describing his new plans, my lips tightened. *Ten days? Thin trails? No map?*

I OPENED GOOGLE Earth and searched for the place names he'd listed. Uaxactun and El Mirador—both in El Petén—looked far off the "Gringo Trail" of popular tourist destinations. I zoomed in. Flat, featureless forest stretched like a green Berber carpet in every direction. I panned around. Other than a handful of brown patches that looked like wetlands to avoid, there was nothing to help guide a hiker. There were no mountains, no rivers, no pastures, no visible roads or trails. It also looked far from Uaxactun to El Mirador—and empty.

I Googled images of El Petén: flat jungle with ancient Mayan buildings, their steps climbing far above adjacent forest trees. A Wikipedia map confirmed that the northern borderlands of El Petén are desolate. Roman's planned route was in the center of the largest area of rainforest left in Central America: seven million acres, covering parts of Guatemala, Mexico, and Belize.

I started picking apart his email, composing a response. He

needed to know that a map wouldn't be much help without topography, rivers, or trails. He needed to be reminded to watch out for the most dangerous snakes in the Americas: the fer-de-lance that kills more people than any other and the bushmaster, an aggressive viper that can grow to ten feet long.

I wrote one email, then another, and another—several. They all said, No, don't do it! or, Do this, it's safer. Each one I deleted, struggling to warn but not discourage him. While he was my son, at twenty-seven he was also his own man, capable, experienced, careful.

> Roman, that way to El Mirador from the west, from Carmelita, looks better. I don't think you should go from Uaxactun. It looks too remote and without a GPS it'll be really hard to know where you are and not get disoriented in the flat karst jungle. Maybe you can find somebody to go with you. It looks like a long ways and it's really remote. Remember that guy we met who was out with a friend in Peru and he got bit by a bushmaster? The friend died before they could get help. I don't think you should go the way you've planned. It seems too dangerous.

I can't send this. He knows what he's up against. He speaks Spanish. He's young. It's his trip and he's there because of me. I've always resented people who warned me off my plans with "No, it's too dangerous" *or* "No, it's impossible" *or* "No, do this instead." *Roman even joined me on many of those trips. How can I deny him his own adventure? Shouldn't I feel satisfaction he's adventuring instead of fear?*

I deleted the discouraging email and wrote instead:

> Be careful with the machete. Clumsy me almost cut off my toe once when it went through the shoe, the sock and into the toe. Also watch for the snakes that sit and wait, motionless, hard to see. Don't want to step on a fer-de-lance or the other big bad aggressive one, bushmaster! And

thanks for thinking of your safety. So is your plan to essentially look for trails heading NW from Uaxactun? Off trail jungle walking can get pretty disorienting.

    Dad

I pushed send and hoped his trip would go well.

What else can I do?

CHAPTER 16

El Petén

Volcano climbing, Guatemala, March 2014.

Roman had dug into the Internet to uncover an ambitious M-shaped route through El Petén. A traveler named Frenchfrog described the route on an online forum: "[It] is almost impossible to do it by yourself unless you have been trained by the Marines or Navy SEALs and you have a perfect knowledge of the jungle, how to find your way." Frenchfrog added, "This is the best adventure I had of all my adventures," but warned, "You must be very careful otherwise you can be easily lost."

In an email to me on the day that he left, Roman described the M's three legs. The easternmost portion started on an unmarked jeep track that weaved below the jungle canopy to Dos Lagunas ranger station. The route then continued northwest on seldom used trails to the most remote Mayan site of all, Naactun (pronounced "Nash-toon"), near the Mexican border. The middle of the M dipped twenty miles to another ancient city, Nakbe, ten miles from El Mirador—Roman's "undeveloped Maya ruin in the jungle." The final leg led to the road at Carmelita, which Roman hoped to reach after dark to avoid its armed tourism cartel. To navigate the M, he would use only a compass, a crude sketch map, and his Spanish.

While Roman was on his traverse, I finished my attic remodel and treated his absence like that of any other adventurer who left me with the responsibility of taking action should they not return by their out-date.

An out-date is the day that we adventurers request a loved one, friend, or some other responsible person to initiate a search to find us, should they not have heard from us by then. In Alaska, this means contacting the Alaska State Troopers, the U.S. Air Force's Pararescue "PJs," the pilot who flew us into the wilderness, or the local mountain rescue team. Besides an out-date, we provide our destinations and route descriptions; the colors of our shelter, pack, raft, and clothing; and any general information that would aid them in a search for us. We strive to be responsible for ourselves and our actions. If we need help, we want our helpers to find us and bring us home.

Roman emailed me his M-shaped route description with an out-date: April 18. If he had not yet emailed me by that day, then I would initiate search and rescue. Tomorrow I am heading to el remate to try and leave details with a gringo guide so you guys have contact info in case I go missing, he wrote. El Remate is the jumping-off point for Tikal. The guide was an older American expat named Lou Simonich.

Ten days later on April 16, two days before his out-date, Roman emailed three sentences: 200km in Guatemala's wildest jungle, only lost for two days. I'll write you more later. Have to find accommodation and wash gear.

It was a relief to hear from him. The full story came the next day. I read it twice, then forwarded it to my own dad and a dozen friends who had watched Roman grow up and done trips with us both. I wanted them to read in his own voice what he had accomplished.

Roman explained why he'd written the six thousand words: There's a lot of stuff I want to record, to see how I remember it right now and how I'll remember it later, and because if I give the brief summary, it sounds not only super badass, but kind of foolhardy. It wasn't really either of those things, just walking for 8 days and asking people where to go.

Before he set off, Roman had spent the night with Lou. A guide and experienced jungle trekker himself, Lou cautioned Roman against telling locals his full plan. Like Umnak's cowboys twenty years before, they would raise an eyebrow at a gringo who took the long way to El Mirador alone. The two stayed up late studying maps and watching Quentin Tarantino movies.

In the morning, Roman helped Lou bake bread. He sharpened his machete and packed away a hand-sketched map of his route. Lou drove Roman to Tikal, where he caught a bus to Uaxactun, a ruin at the end of a dirt road. He camped there for the night. Lou's advice not to mention anything about El Mirador was good, Roman noted, as the reception I got from the locals was to discourage me walking there.

Anticipating the jungle to be hot, and looking to save weight in order to carry water, Roman had left his sleeping bag and extra clothes behind with Lou. Carrying only his jungle clothes and big Mexican pack for insulation, he would sleep cold most nights of his trek. Chilled and stiff hours before dawn, he'd wake to the lion-like roar of howler monkeys, climb out of the backpack he'd pulled to his waist like a bivy sack for warmth, then spark a fire in the dark and hunker over it while boiling the day's drinking water.

He left Uaxactun at dawn following directions he didn't know were bad, checking his compass and sketch map relentlessly as he passed side trail after side trail. Eventually, he left the rolling wet hills of karst and entered dry, flat scrublands with an overstory of short palms. The road straightened out. A gallon into his water reserves he came to his first *aguada*, or water hole, next to a well-used campsite. He stopped there and boiled up a gallon in the afternoon heat.

It might seem ironic that Roman saw the crux of his rainforest crossing as water, but it was the dry season on an enormous limestone shelf, porous as a brick of Swiss cheese. Frenchfrog himself had run out of water mid-trek.

Water wasn't his only concern though. Being alone in a foreign wilderness left him nervous. Traffic on the jeep trail could include narcos, thieves, or belligerent locals, as well as friendly campesinos, archaeologists, rangers, and even tourists. His apprehension increased as a guy with a long gun slung over his shoulder zoomed by on a motorcycle.

I could barely return his bemused greeting, Roman wrote, as my eyes were fixed on the firearm. It was old and rusty and looked like a break-barrel 16-gauge. Not a narco weapon, but a poacher's. He was smiling when he went past, though. Roman relaxed but not for long. As dusk settled, he upset two huge birds in the dark. The pair erupted in honking, clumsy flight, startling him. From his description they sounded like long-necked, long-legged, turkey-looking guans.

On his second morning, he woke cold, boiled water from the aguada, and left before the sun had risen. Despite his heavy load, his feet felt fresh, his spirits high. At dawn, warm, welcome light spilled into the forest—then something crashed out of the brush and onto the trail.

Before I could register that the barking, charging blur was a terrified wild pig and not a ferocious feral dog, my machete was out and pointing in the right direction. Later, a puma stepped into view, looked at him, then loped down the trail and slipped back into the forest. I didn't pull my machete out that time, he said.

Besides the big animals with teeth, there were lesser creatures that could also bite. Roman nearly stepped on a couple of snakes camouflaged among the dry season's leaf litter. One long one—stretched halfway across the jeep track—rattled its tail in the leaves and enticed him as a potential meal. I started looking around for a stick, he wrote, as six feet of snake would make a good dinner. But I think it realized my intentions and slunk off into the underbrush.

Roman continued northwest on a lesser-used track mentioned by Frenchfrog. He hoped to bypass the Dos Lagunas ranger station by heading more directly to Naactun through the jungle. As the side track braided, he applied his volcano-climbing rule and followed the best-used tracks headed toward his destination, confident with his route finding: "Marching merrily along, avoiding bullhorn acacia [a small tree protected by painful stinging ants] and spiny climbing palm, I'm thinking about how well my trip is going, and if this is jungle travel, you sure as hell don't need to be a Navy Seal."

But soon the track thinned and he trail-blazed trees with his machete to ensure his return on what now alternated between game trail, dry creek bed, and impenetrable bracken, a tropical fern that grows in tangled thickets five to ten feet high. It can take hours to make a hundred yards through the stuff.

After fighting his way through one such thicket, he found he'd lost the trail. It was midafternoon. He set down his pack to scout beyond the dry creek in wetter, taller jungle with a head-high understory of dwarf palms. Intrigued by a 150-foot hill, he climbed it but lost his way on the descent: everything looks the same in the jungle, and the forest's multilayered canopy blocks the sun's use as a navigational handrail. Disoriented and a little freaked out by how quickly the uniformly green landscape had swallowed his trail, he was relieved to stumble back to his pack.

Tropical wilderness can be a frightening place alone. Following a bearing by compass or GPS inevitably leads through impassible swamps, tangled vines, and other vegetation hiding poisonous snakes, painful stinging insects, scorpions, and centipedes, or the spines, thorns, and rash-inducing resins of plants. Nights are long. Big cats—and desperate humans—sometimes take the lives of solo travelers.

Camping his third night, the coldest yet, he stuffed plastic bags for insulation into his clothes. Up at three, he waited around a fire for dawn. "I decided that there was no trail to follow, I didn't know what I was doing, and I should go exhaust my options on the jeep road." But instead of turning back, Roman further explored the jungle. He caught a lizard and killed it. In his journal he wrote: "tried eating a lizard. Gross."

He ended up spending about half the day around his campsite looking for a trail that might lead west to Naactun and El Mirador.

There definitely was a trail there: an ancient Mayan road in the forest. It ran for maybe two hundred feet in a perfect line, six feet wide and three feet high—a raised walkway called a *sakbe*—that ended in a ruin excavated by looters. Intrigued by the ancient route, he went deeper into the jungle. He had plenty of water—almost thirty-five pounds worth—but his burden forced him to drop his pack while cutting trail and return to shuttle it onward with ants, spiders, dead twigs, and dried leaves stuck to his sweaty neck and arms.

One of his blazed trees wept white sap below the unmistakable V scars of rubber tappers. Encouraged he might still find a trail, he pushed onward until dusk. He stopped at the edge of a giant limestone sinkhole. Hoping to find water he found "nothing at the bottom but wasps, crumbling limestone mud, rotten logs, and the promise of never being found."

That night, his fourth in El Petén, he learned that if he draped his plastic tarp over his mosquito-net tent he stayed warmer and slept well. Fire was tough to start in the dewy morning with his candle wax, but he found that the flaky, flammable bark from the "tourist tree" (so-named because it looks like a peeling, sunburned tourist) brought the fire to life.

"The next day I went exploring for four hours before deciding it was time to go back and find Dos Lagunas. The jungle was making me claustrophobic and I wasn't sure how well I could follow my trail back out." He discovered it was much easier to walk out than walk in: a trail that took two days to hack with a machete took just three hours to the jeep track.

Roman strolled into the Dos Lagunas ranger station just before sunset. There he found "an old white guy" wandering around and "four somewhat standoffish, if curious rangers. On a whim, I told them I was going to El Mirador. They said I could camp there."

Pleased to be welcome and happy to have company after five days in the jungle alone, Roman dropped his big pack, wiped the sweat and dirt from his face, took a long draw from his water bottle, and pulled on a dry shirt, wondering what the other gringo was up to so far from the end of the road.

CHAPTER 17

Finding Carmelita

Campfire in Mexico, January 2014.

A young ranger grilled Roman on where he was from, probably since the "old white guy" was a paunchy, middle-aged Russian without any Spanish. The rangers hoped Roman could communicate with him. But he did little better than they, learning only that the Russian restored museum paintings in St. Petersburg and hoped to make it to El Mirador by way of Nakbe.

In Spanish, Roman told the rangers what he had done and what he wanted to do. He showed them his sketch map and compass. In response, they gave him a recent, color map of the route between Dos Lagunas and Naactun. Over a shared dinner of beans, tortillas, and Nescafé, they told Roman that the rangers at Naactun would have better information about getting from Naactun to Nakbe. They asked if he wouldn't mind traveling with the Russian. Roman explained he didn't have enough extra food for the Russian's slow pace. Knowing it was safer to travel together than alone, the rangers offered some of their food to take. Roman accepted a package of ramen noodles and said he'd watch out for the Russian by marking trail and leaving water.

Off by seven the next morning, Roman left the rangers Q50 (about US $6) and two packets of cookies. Guatemalans tend to be generous and he liked to reciprocate. He had discovered that he was often in someone's home, not just their place of business, "so I try and remember the manners Mom taught me. Mostly washing dishes voluntarily. That has worked well for me my whole life, everywhere."

Two hours out of Dos Lagunas, the head ranger rolled up on his dirt bike to ask if Roman wanted a ride. "Older Guatemalan men I've met, the ones that seem like old cowboys, tend to be very warm and fatherly. The *jefe* and other older ranger were no exception. Concerned, understanding, helpful, interested, with a twinkle in their eye and a knowing smile about 'aventura.' "

Roman accepted the jefe's offer and hopped on his dirt bike. An overloaded ATV carrying two more rangers plus the Russian, and a fourth ranger on another motorcycle, followed. "It was fun," Roman wrote, "but also probably the most dangerous thing I've done here. I constantly had to dodge vines, scoot back on the seat to reduce my profile so my knees wouldn't clip trees, and brace with my arms while lifting my feet up to avoid the sides of the deep ruts we'd sometimes fall into."

An hour and a half later, Roman had completed the first leg of the M to the central crossroads of the ancient Maya: Naactun.

The site has a higher density of ancient sakbe walkways than anywhere else in Mesoamerica. While there, he met a team of archaeologists led by a Guatemalan named Carlos Morales-Aguilar, a preeminent researcher in El Petén. Morales-Aguilar enthused over the significance of Naactun, the center of civilization for ten million Maya a thousand years ago. Roman spent hours wandering around the excavations and ruins. With a new map sketched in his notebook, he headed for Nakbe, the middle vertex of the M.

This leg in the remote heart of El Petén is rarely traveled. The trail grew faint and braided. Unsure where he was, he at least knew how to get back. For days he'd been learning to differentiate machete scars from natural damage to trees, and vehicle damage to roots from horseshoe damage. He'd learned to tell if a poacher's dirt bike had been down a dry, hard-packed trail based on the patterns of broken termite tunnels on the ground.

Sussing out the tangle of footpaths and old ATV trails with his compass and newly sketched map, Roman left trail-blazes for the Russian as he hiked. That night, his sixth since leaving Lou's place in El Ramate, he set up camp under fragrant lemon and grapefruit trees next to a large aguada. An exquisitely excavated Mayan wall stood nearby, sculpted in angel-like wings and other human and nonhuman forms. The dry season is terrible for ticks and chiggers in Central America, and Roman spent an hour that night picking off parasites. The itchy rash on his feet, he realized, was not a rash at all, but dozens of tiny ticks, each raising an angry little welt. And they weren't just on his feet. They were everywhere: ankles, arms, crotch, armpits, belly. "I took a small bath in DEET," a potent insect repellent developed by the military, "which killed them. They were easy to scratch off."

Up at six to boil water for the day, he pondered his situation. If he was camping at La Muralla, halfway from Naachtun, he would reach Nakbe by noon. But he couldn't be sure he had been on the right trail and considered the risks that he faced. "Worst case scenario I got bit by a snake and died slowly. Not much I could do about that. Second worst was wandering too deep into the forest, got lost, and couldn't find water."

He noted that by staying on the "cleanest trails and leaving blazes, and never venturing more than two days from an aguada" he could avoid getting lost and not running out of water. He also recognized there were wild fruits, like the sweet chicozapote, and plenty of snakes and lizards. "I think I could have foraged a meal every day just walking, if I didn't mind skinny grilled lizards."

He also worried that he had misled the Russian behind him. He left two quarts of boiled water for the paunchy painter, then pushed on, arriving at Nakbe in time for lunch. He shared his fresh limes and grapefruits with the rangers. They were astonished with his trek, but wouldn't have done what he'd done, not alone: too dangerous, they said. His story of the Russian intrigued and humored them, but worried them, too. They decided to head back to La Muralla to find him.

During his few hours at Nakbe, Roman toured the ruins. From the top of its major temple, he could see the partially cleared pyramid of El Mirador rising 250 feet above the flat expanse of jungle that reached to the horizon. It looked distant, but in little more than two hours of fast walking he'd be there.

One of the rangers, Miguel, needed to make a supply run to El Mirador and invited Roman along. The ranger, who carried only an empty pack, was pleased with their rapid time and surprised that Roman—carrying a big pack—had kept up. Miguel's pace left Roman dehydrated and hot, with big blisters on his toes and heels. "Oh well," he wrote in his journal. "Only one more day."

At El Mirador, Miguel shared Roman's story with a cook who

offered him a dinner of "beans, tortillas, and some delicious scrambled egg dish. I gave her the rest of my limes. The rangers at El Mirador were also interested to hear about the Russian, and had a laugh."

It was good to see that Roman interacted with the people at every stop along the way and even better to find that he shared what he could with them: he exercised good wilderness etiquette. I was delighted that he cared for the Russian, too, whom he didn't know, but realized needed his help.

On his last day, with a late start at eight, he took off to cover the final thirty miles to Carmelita. Half an hour out, a helicopter flew overhead carrying Richard Hansen, the famed researcher who'd put El Mirador on the map in the eighties. "It would have been cool to meet him. Maybe I should have stayed another day. But who knows? He's probably sick of tourists. He had all semester to be asked stupid questions by undergrads."

Roman felt the effects of the previous day's race pace. The hours dragged on. His feet hurt and his chigger bites itched as he pounded the hard-packed trail. To ease his tender feet, he took softer, parallel paths. By nightfall he was out of water, thirsty and spent. Worse—because time and distance stretch in the dark and he hadn't yet reached Carmelita—hobbling on sore feet through the night he worried he had taken a wrong turn in the dark.

While he was sure he was nearing Carmelita, he heard only what he described as "the primordial hoots and howls of the New World tropics." Just as he was considering the prospect of a dry camp and calling it a very long day, he heard people. He followed their voices to a house and asked where Carmelita was. They laughed: he was *in* Carmelita. The husband led Roman to a tourist hotel where he quenched his thirst with bottles of soda, water, and Gatorade, and bought a bar of soap. "I took my last DEET bath, a mondi, then slept for the first time in 9 days without being cold. It rained that night, hard. I was glad I hadn't stayed another night out. My tent would be miserable in a rainstorm, and the trail an abominable mud pit. I did worry a little about my Russian, though."

At four the next morning, even after the marathon efforts of the last two days, he caught the "chicken bus," one of the colorful local transports packed with people, arriving at Santa Elena six hours later. He spent the day washing clothes, limping, eating, and writing the story of his journey.

Few people have done El Petén's M route. Fewer still have done it alone. Roman had proved himself in Central America's biggest wilderness. I was impressed—also relieved.

After Guatemala, Roman visited Belize. "The only people I've talked to that liked Belize were the young European women that like everything, especially poor people, or white girls who smoke too much and don't take care of their hair." I chuckled at Roman's distaste for "hippies," whom he saw doing little more than drugs while lounging around their hostels. He headed south to Utila, Honduras. There, he paid $289 for an advanced diver certification, a lot of money, he said, but worth it. He was the only student in a course that included accommodation, gear, and seven dives on which he swam with whale sharks and made night dives.

A month after El Petén, Roman emailed plans and a map of eastern Honduras. Again, he lamented leaving his packraft behind. He envisioned a three-hundred-mile river trip down the Patuca River through the heart of La Moskitia, second only to El Petén as Central America's biggest roadless area. The Patuca itself ends at the famed Mosquito Coast, shared by Honduras and Nicaragua.

Roman described his planned trip to his college friend Brad as "400 miles of jungle swamp without good maps through North America's cocaine hub in the murder capital of the world." He hadn't shared that reputation with me. He wrote me that he was headed to El Salvador to look for a canoe for his river trip through La Moskitia, which I knew only for its biodiversity values, not its lawlessness.

If he had told me, I would have likely written him that lawless humans are more dangerous, more unpredictable than wilderness. Once a criminal breaks one law—like smuggling drugs— it's easier to break another—like robbery or even murder. Risk management of mountain, river, and wild animal hazards is more straightforward than planning for outlaws. But after his El Petén trek, it was clear he could take care of himself. It sounded like he was ready for another full-bodied adventure and I looked forward to the stories he'd tell.

CHAPTER 18

South to Costa Rica

Scuba diving the Bay Islands, Honduras, May 2014.

While in El Salvador looking for a canoe, Roman met Jeremy, a Canadian who was also interested in La Moskitia. Because they had found only sit-on-top kayaks, unsuited for their trip, the two decided they would rely on local transport instead. They headed for Palestina on the banks of the Patuca River in Honduras and boarded a sixty-foot cargo canoe, loaded with hundred-pound bags of rice, cases of soda, and leaky fifty-gallon barrels of gas. "All the necessities of village life," Roman noted.

The boatman did his best to steer the canoe down the low, dry-season river, but the current pinned the overloaded craft on the rocks of a shallow rapid. The captain ordered the fuel drums jettisoned and the boat slipped off the rocks and headed downstream, where Jeremy and Roman helped retrieve the barrels.

Unlike the rest of Honduras, La Moskitia is indigenous, not Latino, and as they headed deeper into the region fewer people spoke Spanish. Eventually they heard only Moskito. At each passing village the big dugout canoe dropped off cargo and occasionally picked up new passengers. Stopped at a gold miner's camp one night, Roman loaned his bug-net tent and tarp to a young Moskito couple. This earned him the respect of the boat owner, who later invited him and Jeremy to sleep at his house. Days later and downstream, the pair found lodging with an evangelical Moskito preacher whose amplified nighttime hymns and sermons drained the power from his little chapel's lights. They had to wait for several days until a boat headed out; a gasoline shortage had stopped all river traffic.

From the start, Jeremy and Roman marveled at the display of weaponry along the Patuca. The teenage kid in the bow of the cargo boat had protected his .50-caliber Desert Eagle from a tropical downpour by stashing the hard-hitting piece in his backpack. Cowboys in tall hats tucked pistols in their waistbands. Honduran soldiers in jackboots and stocking caps brandished submachine guns, assault rifles, and side arms. In one village, Roman watched a shirtless man with an automatic pistol sandwiched between his brown belly flesh and pants trade gold flakes for Doritos and Pepsi.

It was hard to tell the narco-traffickers from the citizens who just wanted protection. Jeremy asked the canoe captain: "Why is everyone armed? Is it dangerous?"

"No, no," the captain replied. "It's quite safe out here. Everyone has guns!"

Everyone has guns because La Moskitia's lagoons, wetlands, and rivers offer Colombian smugglers a place to refuel and hide out on their way to land routes through Guatemala and Mexico to the U.S. cocaine markets. Small, open boats with multiple outboard engines ferry the drug from its origin to landfall in eastern Honduras.

Jeremy and Roman talked their way onto another big dugout canoe with an outboard. This one cruised down the wide river in the dark. Hurtling through one of the most active cocaine transit areas in North America, the two ate cookies and looked at the stars as the captain texted, motoring at full throttle. The next day they boarded a twin-engine jet boat full of passengers. The boat followed narrow canoe trails through a swamp at high speed, then careened wide open around corners through the lagoons of the Mosquito Coast of the Caribbean Sea.

As I read Roman's account, I could picture Jeremy and Roman looking at each other and grinning, shaking their heads during a "thrilling Disney World ride. Except the overhanging vines were actually close enough to hurt and the oncoming boat, also going at full throttle, came very close to being very bad."

Arriving at Puerto Lempira, the largest town in La Moskitia, they looked for accommodation. It sounded grim: "100 lempira [$10] a night got you a soiled mattress, bug bites, and used condoms under the bed. 50 lempira was either malaria or getting stabbed in an alley." A hospitable guy named Junior charged them one hundred lempira for the only clean place in town. Over beers, Junior barbecued chicken and cooked up a Honduran specialty of cheese over beans in a traditional clay pot. In the morning, they toured Puerto Lempira, where Junior pointed out the narco-traffickers' kids, their bodyguards, even who'd been shot, how many times, and by what caliber bullet.

From Puerto Lempira on the Mosquito Coast of the Caribbean, they caught a pickup truck to Nicaragua. The dirt road passed "through really beautiful country. I'm not sure why I liked it so much." The landscape reminded Roman of a surreal Dr. Seuss version of Alaska's arctic tundra: "Too even, too green, too smooth, too pretty," he said, "to be quite right." Safely past multiple military checkpoints, they arrived in Nicaragua without passport stamps. With indigenous La Moskitia behind them, they had returned to Latin America.

THROUGH LATE JUNE into July, Roman headed south. I sensed from his emails that he was homesick after eight months away from Alaska. He surfed in Nicaragua for two weeks, joking that packrafting and surfing have almost no crossover skills, except maybe swimming. Worried about rabies, he asked us what he should do about a street dog bite to his leg that drew blood; gave a Honduran recipe to Peggy; recommended we watch BBC's *Sherlock*; and, If you dont have it already, he suggested, you should get New Order's 1987 Substance Album.

His music suggestion recalled that sweet spot in his adolescence, between boyhood and manhood, when he saw me as both fun and cool. During that golden age, we shared music and books, swapped interests and insights. And as he grew to know more than I did in economics, genetics, and politics, he shared his knowledge, enriching my life. It was during those years that we stared at thousands of bugs and reminisced about Borneo, packrafted whitewater when no one else did, and discovered he could beat me at chess.

"Where in Costa Rica did we go with the APU class?" he asked in mid-June.

During the month of January 1999, I led a dozen students from Alaska Pacific University on a tropical ecology course to Costa Rica. Roman joined the APU class as a precocious eleven-soon-to-be-twelve-year-old. We crossed the small nation from coast to coast in a little chartered bus and studied Central American ecology en route. We saw poison dart frogs on the Caribbean side, ctenosaurs and crocodiles on the Pacific, and rafted whitewater in between. We walked for a week across Corcovado, through its lowland rainforest and along its beaches on the park's most iconic backcountry route. Independent travel was possible in Costa Rican national parks then and we walked on and off-trail at will. At one point we forded a lagoon on an incoming tide said to carry sharks in and crocodiles out. Young Roman waded nearly to his neck.

Fifteen years later, Roman was collecting volcanoes, high points, and major jungles throughout Mexico and Central America. He'd visited the Lacandon Jungle in Mexico, El Petén in Guatemala, Belize's Maya Mountains rainforests, and La Moskitia in Honduras: Corcovado National Park on Costa Rica's Osa Peninsula and Panama's Darién Gap were all he had left.

He told some of his friends that Corcovado would be training for the Darién Gap, a literal gap in the transcontinental road system between Panama and Colombia. Because it is occupied by militarized Panamanian and Colombian border police, paramilitary revolutionaries, and drug traffickers (not to mention fer-de-lances, bushmasters, other poisonous snakes, bullet ants,

dengue, malaria, and more), the Gap is one of the most dangerous places on Earth.

Roman emailed me June 6: I've spent the last week or so trying to
figure out how to do the Darien gap and it's starting to give me bad dreams.
He wrote his college friend Brad:

> Seriously planning a trip through the Darien Gap. It's fucking stupid
> and there's a really good chance I die or get kidnapped. Senafront, the
> Panamanian border police, doesn't let foreigners cross the Colombian
> border over land. My plan is to get permission to go to Darien National
> Park, dip out on the rangers, follow a river south up into the low,
> extremely steep limestone border range, cross it into Colombia, then
> follow a river down to an Indian village and hire a boat out.

I shared Roman's sentiments about the Darién. It sounded too
dangerous to try. But while part of me hoped that he wouldn't,
another part hoped that he would. In my younger days, like many
adventurers, I had imagined its wilderness as a worthy challenge
to travel. But its social hazards of lawlessness and paramilitary
groups had made it too dangerous for me. If we as parents live
vicariously through our kids, then after Roman crossed the
Darién, my empty ambition to try wouldn't matter. On the other
hand, I knew well its reputation.

On the Fourth of July 2014, and still in Nicaragua, Roman
asked me in an email, Do you have any super-secret access to topo maps
of central american countries?

I wish, I wrote back. Try googling ESRI world topo. Better than nothing. I
checked the world topo's version of Corcovado to compare it to
somewhere we'd been. It identified the Osa Peninsula as part of
the canton of Golfito.

On July 6, Roman arrived in San José, the capital of Costa
Rica, and bought a backpack that he planned to use in Corcovado

and farther south. Roman's bulky Mexican pack held his insulated pullover, a thin summer-weight sleeping bag, two stoves, and our old Kelty tent. He also carried Forrest McCarthy's small yellow duffel as a daypack. On Tuesday, July 8 at eight in the morning, Roman left San José on an eight-hour bus ride headed for the Osa Peninsula.

His destination: Corcovado National Park.

CHAPTER 19

"The Best Map Yet"

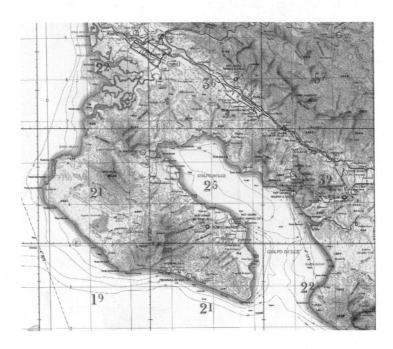

The Osa Peninsula and Golfo Dulce, Costa Rica.

The Osa Peninsula, situated just north of Panama's west coast, separates the muscular Pacific from the calm Golfo Dulce, Spanish for "Sweet Gulf." The Osa's main road is a two-lane highway that parallels Golfo Dulce as far as Puerto Jiménez. There the pavement ends. In the nineties, the Osa's remoteness, abundant wildlife, and sparse population drew a cohort of North Americans and Europeans who settled and started businesses that thrived until the recession of 2008.

Today, a collage of billboards greets travelers at the end of the highway promising "waterfalls, tours, and massage" and "Affordable Beachfront Luxury!" Also available: "sport fishing," "sea kayaking," even "zip-lining." These tourist establishments are small, family-run operations that contribute to the economy and English fluency of the locals, but hardly qualify the Osa as a tourist mecca like those farther north.

A sleepy town with colonial roots dating back to the mid-1800s, Puerto Jiménez's economy evolved from banana farms to gold mines. Its commercial district spans six blocks where dogs lie in the street and free-range chickens scratch in the dirt. Scarlet macaws—red, yellow, and blue–colored parrots the size of ravens—squawk overhead. There's a hospital, a police station, an office for Cruz Roja (Costa Rican Red Cross). A Catholic church fills a city block–sized campus. Young men play soccer at a fenced-in field on the edge of town.

There is one gas station, two banks, a *farmacia*, maybe five bars, two supermarkets, and a hardware store. For tourists, there's a surf shop, a handful of restaurants with English menus, shops with colorful toucans carved from wood, tour centers with grease boards announcing the day's activities, and hostels that eagerly invite backpackers.

A block away, the gentle Golfo Dulce laps at a sandy beach that opens into a channel lined with mangroves where an occasional heron patiently fishes. Around the point, beginner surfers try their hand at a left-hand break. Beyond town, the dirt road bumps along for forty-five minutes to Matapalo, a diffuse beach community centered on the best surfing on the Osa. Another forty-five minutes past Matapalo, the road ends alongside a long paved airstrip at the village of Carate, on the opposite side of the Osa from Puerto Jiménez. Beyond is the most remote beach in Costa Rica, twelve short minutes from Panama by small plane.

On any given day, Ticos and Ticas—local Costa Ricans—

saunter down Puerto Jiménez's sidewalks, where they stop and gossip in greeting. Old cars and dusty SUVs pass each other with inches to spare. The ancestors of Puerto Jiménez included pirates and Indians, convicts and civil war rebels, squatters, gold miners, crocodile hunters, banana farmers, cattle ranchers, and those who fled San José's crime and Nicaragua's revolution.

Tourists—especially clean-cut, evenly tanned, lithe young men and women wearing flip-flops, tank tops, and sun hats—sit at open-air restaurants and page through their Lonely Planet guidebooks. Most come to the Osa to stay at an eco-lodge or to visit Corcovado National Park, considered the country's crown jewel of conservation. Large by Central American standards, the park sprawls across 100,000 acres with big Amazonian animals to match: jaguars, tapirs, harpy eagles, crocodiles, bushmasters.

The Osa's forests also hide poachers, illegal miners, drug smugglers, and murderers. At the north end of the peninsula in Sierpe, a small cowboy town in a mangrove estuary, two tons of cocaine were found in an underground cellar in 2016. In 2011, two North American women in their fifties were found murdered near Puerto Jiménez in separate incidents. In 2009, two Austrians in their sixties went missing from their blood-spattered house in Dos Brazos, a twenty-minute drive from Puerto Jiménez. Even Olaf Wessberg—the ex-pat Swede considered the father of Corcovado National Park—was murdered in the jungle near Puerto Jiménez in 1975.

WHEN ROMAN STEPPED off the long-distance bus into Puerto Jiménez's sweltering heat on the afternoon of July 8, he knew none of this history of violence lurking just beyond the tourist billboards and hostels. But the Osa's confluence of wild nature and dangerous people is not rare in Central America; nor was it new to Roman.

Sometime after four that afternoon, he checked into a place that his Lonely Planet guidebook called Cabinas the Corner Hostel. He wrote his name and passport number in the register.

On July 9, from an Internet café a block from his hostel, Roman emailed a friend: Currently on the Osa Peninsula on the Pacific, right next to Panama. There's a national park I am going to sneak into and bushwhack around in. Practice for the Darien. He went on to say that he might cut his trip short. He needed to buy his ticket home, where he'd need an apartment, a car, a job, and pay for another semester of school. Costa Rica is burning through my cash. Otherwise, I wanted to see Colombia and climb some mountains and go trekking. I think South America is going to wait for another trip.

The same day, he sent detailed plans to Peggy and me in two emails. The first at 9:02 AM said he was in Puerto Jiménez shopping for food to head into Corcovado. Five months before Roman arrived on the Osa, in February 2014, Corcovado National Park had enacted new regulations that required that all visitors who enter the park have a licensed guide. Roman had spent less than twelve hundred dollars a month since January. Even if he could fit a guide into his tight budget, he neither needed nor wanted one: Anyway, Im heading in offtrail tomorrow, just west of the Los Patos to Sirena trail. Its about 20km, then Ill hit the coast and follow the Madrigal trail out at night. I am going to try to follow the Rio David south, then hop over to the Rio Claro. . . . I anticipate the highlands to be slow and wet.

The highlands—where a maze of poacher and peccary trails crisscross a summit plateau cut by shallow canyons called Las Quebraditas—are indeed slow and wet. The plateau, officially off-limits to all but park guards, is notorious among miners and rangers alike as a disorienting landscape of rainy bamboo forests tangled in vines.

I am not sure how long it will take me, but Im planning on doing 4 days in the jungle and a day to walk out. 5km a day is an abysmal pace, but it's hard to keep a straightline without a horizon. Ill be bounded by a trail to the west and the coast everywhere else, so it should be difficult to get lost forever.

Those final two words would haunt me for years.

Twenty minutes later, at 9:26 AM, he sent a link to the map that he would carry with him. Ok, I found what seems to be the best map yet. Ive been looking at a variety of other maps with rivers and trails in different places, with different names. He described a new plan: Im going to try and follow the Rio Conte up, then head south to Rio Claro, which he would follow to the coast and out to Carate. Its supposed to be the rainy season, so I dont know how passable these hills are. You know how steep and slippery this kind of terrain can be.

Then, of the $3,436 in his bank account, he withdrew 50,000 colónes—about $95—from an ATM a few blocks from his hostel. Across the street at the supermarket, he bought five days of food for just over $25. He cooked his dinner in the hostel's kitchen, then spread his gear on his dormitory bed and divided it among his small yellow duffel bag, his big Mexican backpack, and his new pack.

Into the yellow bag he put his Lonely Planet guidebook, a spiral notebook, beach supplies, and clothes. Into his big Mexican backpack, he stored his Kelty tent, sleeping bag, Jetboil stove, puffy jacket, and other warm clothes he had used for climbing volcanos; his flip-flops, blue jeans, and belt; plus other clothes and another notebook. For his five-day trip into Corcovado, Roman filled his new pack with cooking and camping gear, food, a machete, topo map, compass, sleep clothes, Visqueen tarp, and mosquito-net tent.

On the morning of July 10 at the Cabinas Corners, he paid $20 to the little old lady who ran the hostel for his two nights spent in the dormitory and another $10 to reserve a bed for his return. He left the big Mexican pack and yellow bag in storage. At around noon, he crossed the street and caught a colectivo for $5 to Dos Brazos, a small village twenty minutes from Puerto Jiménez and located on the mountainous edge of Corcovado National Park.

Roman wasn't headed for the Rio Conte after all—but told no one his new plans.

Dos Brazos means "two arms" in Spanish, referring to the two arms, or forks, of the Rio Tigre that come together there. The village's three hundred miners, subsistence farmers, and their families live in simple homes along two short gravel roads, one along each river arm. At the junction is a *pulperia*, one of many small wooden shacks with sheet metal roofs that are sprinkled across the Osa. They sell snacks, drinks, and newspapers. This one sometimes buys gold from local miners.

Early in the afternoon of July 10, Roman climbed out of a colectivo across from the pulperia, shouldered his pack, and headed alone up the right arm of the Rio Tigre—El Tigre— into the jungle of Corcovado.

PART III

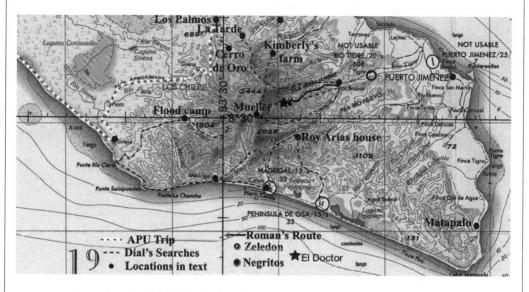

The southern half of the Osa Peninsula, Costa Rica. The distance between grid lines is six miles. The black star is the site of discovery. N.B. "Los Palmos" on north edge should read "Los Patos."

CHAPTER 20

"email, please!"

Paro Takstang Monestary, Bhutan, 2012.

While Roman was exploring the cultures, mountains, and jungles of Central America, I was finishing up home projects and making short day trips and planning a long packrafting trip in the nearby Talkeetna Mountains. I enjoyed hearing about Roman's trips via email, but looked forward to having him home. When he'd written that he'd been bitten by a dog in Nicaragua and worried it had rabies, I'd even thought to ask him if maybe it was time to come back. But I didn't.

It had been gratifying to me as his father to see him out on his own. He would return world-wise and confident with a broader view of life. His Spanish would be excellent. His view on economics and the role of the United States in Latin America would be better informed. It was also clear that his adventures had grown naturally from his upbringing: our family trips to Australia, Borneo, Alaska's wilderness, and elsewhere. I wanted to hear his stories, perspectives, and insights firsthand.

On July 14, home from a Talkeetna Mountains packrafting trip with my friend Gordy Vernon, I scanned Roman's last email: OK, I found what seems to be the best map yet. Unpacking and catching up, I read no further. But buried in the thread—unseen for another week—was his email that said he was planning on doing 4 days in the jungle and a day to walk out. We'd been emailing about super-secret topo maps of Central America. The two threaded emails seemed part of that conversation. I didn't read past the best map yet. If I had, then I would have known he planned to be out from his Corcovado trip the very next day.

July 15 was his out-date.

The summer of 2014 was sunny in Anchorage and Peggy and I kept busy. We worked on house projects until peak salmon season, then drove to the Kenai Peninsula to dip-net fish for our freezer. We camped on the beach where the milky-blue Kenai River slides into the glacier-gray Cook Inlet and the sea breeze keeps July's mosquitoes at bay. Beneath a clear sky and sunshine, we enjoyed the views of mountains rising above fishing boats plowing back to port, their holds full of freshly caught sockeye salmon. The reds were running strong and people lined up shoulder to shoulder, standing in the river, their long-handled nets straining against the current as they excitedly pulled in fish when they felt a gentle bump in their net. We saw friends and filled our coolers with shiny sockeyes.

Still, it nagged at us that we hadn't yet heard from Roman. I checked my phone for new emails as often as the spotty coverage on the Kenai allowed. *Nothing.* It had been six months since I'd seen him. He hadn't told me exactly when he would be back from Latin America, but I hoped to have him home soon. I missed him.

Peggy and I returned from fishing on July 18, cleaned the twenty salmon we'd caught, and set to work finishing a siding project on our house. Days crept by. Still no word. We weren't alarmed, just a bit surprised. Hardly a fortnight would pass since Veracruz when we wouldn't hear something from Roman. On July 21—twelve days after he had last written—I sent a gentle reminder: "Let me know when you get out." His email linked to the one starting with the best map yet sat unread in my inbox.

On July 23, Peggy and I wandered between fasteners and paint at Lowes, wondering aloud to each other why we had heard nothing from Roman. Two weeks had passed. The longest stretch he'd gone without contacting us after Veracruz had been ten days, during his trips across El Petén and La Moskitia. We were worried now.

"I need to look at his last email again," I told Peggy. "I didn't really read it carefully and I'm not sure what he wrote. It seems like it was just about maps."

Then and there in Lowes Home Improvement, Peggy felt nauseous. We left empty-handed to drive straight home and read his emails carefully. I opened the July 9 thread where the words heading in off-trail tomorrow . . . 4 days in the jungle and a day to walk out spilled across my screen. My face went numb.

OH NO! He's way overdue—fuck!

I should have been paying closer attention!

Shock washed over me. Then guilt. Guilt over the fact that I hadn't read his email thoroughly, that I hadn't given him the attention he deserved. That, maybe, like Peggy pointed out in nearly every argument, I spent too much time on my own trips, on my own interests.

"Peggy. This email says he should have been done, like"—I struggled with the arithmetic—"like, ten days ago! *Something's wrong!*" I turned to her. Her forehead tightened, cheeks slack. She saw my terror; it increased her own.

WE JUMPED INTO action. She slid me a notebook and pen across the table, then got on the phone and called Jazz. I set to work on the computer, my hands shaking. Fighting panic and rising nausea, I googled Corcovado national park guides, looking for someone to help us.

My Spanish too poor to call, I shot off an email to Osa Corcovado Tours.

> My name is Roman Dial and my son, Cody Roman Dial, age 27, is missing in Corcovado National Park. He is about 177 cm tall (5 feet 10 inches), with blue eyes, brown hair and glasses. He weighs about 63 kg (140 lbs). He should have a blue two-person tent.
>
> He has been traveling for several months in Central America and doing treks in the jungle, always without a guide.
>
> He emailed us on 9 July and said that he was heading into Corcovado National Park on 10 July for five days alone. He should have returned ten days ago, and he always reports back to us. But we have heard nothing and now are worried.
>
> He wrote that he would be hiking off-trail to the east of the Los Patos to Sirena Trail. He said he'd be walking about 5 km a day for 20 km off trail, following the Rio Conte up, then crossing the mountains over to the Rio Claro and follow that to the coast.
>
> Again he said he would be gone for 5 days and that was almost 14 days ago. Can you please advise me what I can do or how we might

look for him? I do not speak Spanish, but perhaps I could call someone and speak on the phone? Attached is a photo from two years ago.

The first picture of Roman I found was from Bhutan. Smiling at the camera, he's a little pudgy with a bit of beard, short hair, and wire-rimmed glasses, wearing a blue shirt. My arm is around him, hand on his shoulder. I attached the photo and the best map yet and hit send.

I bought an airplane ticket to leave the next day for Costa Rica. I could not stay in Alaska. I would not leave the search up to others. He was my son. My responsibility was to him. Part of the Alaskan creed is that we take care of our own. I had been on enough rescues to know our system worked. Roman had sent me his plans and a map because he knew that if something happened to him, I would come get him.

I had introduced him to the tropics, to wilderness, to world travel. No one knew better what Roman might do. But I needed experienced, reliable help we could trust. I called Gordy. A world traveler himself, he once lost six fingertips when he quit his own attempt at the summit of Mount Everest to rescue another climber on the mountain. He had also lost his father and two siblings in a tragic airplane accident.

Gordy went silent for a minute when I told him the news. He'd been on the Grand Canyon trip with Roman and me. He appreciated Roman's toughness, wit, and modesty.

Gordy's voice was slow and measured, fighting back emotion. "Nah, Roman, my Spanish just isn't good enough for something like that. You'll be better off with Thai." Thai Verzone, his Wilderness Classic partner and protégé, had been both a Latin American studies major in college and a mountain guide in Peru, Ecuador, and Bolivia. He speaks Spanish fluently.

Gordy went on. "You know, what I would do is get hold of Roman's bank records. Those might say a lot about where he was, and where he was going." This advice from a close friend helped. Peggy would try over the coming days, but it took years in the end to get the records.

I called Thai. "Thai: Roman's missing in Costa Rica."

"*What?*"

"Yeah, he wrote he'd be gone on a five-day trip in Corcovado, but he's like ten days overdue!"

"*Oh shit—ten days!*"

"Listen, can you go down there with me? I need you. I'm leaving tomorrow and could really use your Spanish and jungle skills."

Thai's wife, Ana, had just had their baby, Maia, three months before. Thai helped Ana at home and worked at the hospital.

Peggy knew how useful Thai would be with language, wilderness, and people. She quickly volunteered: "I can watch Maia for Ana while Thai goes with you."

I relayed this to Thai. "Peggy says she can help Ana with Maia if you can come."

"Let me check with Ana and the clinic, but I'm pretty sure I can do it. How long will we go for?" Thai had his own life.

"If you could come down for ten days, that'd be great. Thai, I really need your help."

Panic inched up my gorge. I choked it down. *Calmness thinks clearly.*

I was terrified that Roman, lost and broken in the jungle, waited for me to come get him. *Hadn't he given me very explicit directions and a map, after all?*

I called the U.S. Embassy in San José, worried it might be closed. A recording said, "Push two for life and death."

I pushed two. A voice answered and said something about a

duty officer, then gave me a Mr. Zagursky's email. I scribbled it in my notebook, then emailed the photo, map, and information to him. I found an email address for the Puerto Jiménez police and sent them the same content, adding Gordy's suggestion to access Roman's bank records. I told them all that I was coming down.

My body crawled with anxiety and a sense of panic held barely at bay. I wanted to be down there *right now*. Every minute counted. While the tropics might seem hot and idyllic, the rains are cold and the chance of rapid infection is real.

I called my boss at work: "Roman's missing."

Her response was immediate, empathetic. "Oh, Roman," she said genuinely, "I am so sorry," as if he were already dead, that I'd already lost him.

Hurt and angry, I told her, "I'm going down to find him and am not sure when I'll be back." What I meant was that he wasn't dead, that she didn't need to be sorry because I would bring him home alive.

THAT EVENING I packed jungle gear. Shoes and shirts and pants and a pack. Compass and head lamp. Stove and a cookpot. Dehydrated food. Bug-net tent and tarp. Sleeping pad and sheet. We would have to move fast. *Bring only necessities.*

My feelings of shock ebbed, exposing a reef of guilt. He'd written that he'd be out on the fifteenth. *I was home then. I should have read his email.*

I should have given him twenty-four hours, then called Costa Rica on the sixteenth to say he was twenty-four hours overdue, then flown there on the seventeenth. *I could have done that.*

But I didn't. A full week had passed since I could have flown down. It was impossible not to see him suffering, waiting, wondering, *Dad, where are you? I told you where I went. I said I'd be out in five days. Dad, come get me!*

Hoping for the best, I emailed him: i am coming down to look for you. The subject read email please!

My flight left for Atlanta at eight-thirty at night on Thursday, July 24. All day I switched from phone to computer, scrambling to put things together. My brain struggled to function as if nothing were wrong while my heart wrested to take control and panic. Peggy, too, called and emailed friends and family, sounding the alarm. Within twenty-four hours, friends set up a fund and deposited money for our search.

The Tico Times, a Costa Rican English-language newspaper, ran a story. People reached out to help. Then, Facebook kicked in. Someone posted on an Osa-specific page about a sighting. I messaged him and he wrote back:

> I am 90% sure that I saw your son based on his picture—did he have a tan safari type outfit (shorts and shirt matching and a hat)? I remember seeing him walking alone along the road and I took him for one of the many volunteers who are always in that area and who never want a ride. I made eye contact with him and he nodded. He was looking into the woods at something that caught his attention. If you want you can call. Hopefully he is simply walking through some tough terrain out in the park and working his way back.

I ached for it to be true. But it couldn't be Roman dressed in safari garb, turning down a ride on a road. I knew that it wasn't. Together we had spent too many months over too many years in too many countries on too many continents for that to be the son I raised.

He was in trouble. I knew.

CHAPTER 21

Dondee

Dondee, MINAE headquarters, July 25, 2014.

Thai Verzone is a good friend and my first choice for trips need-ing multiple skills. Son of an Italian father and a Vietnamese mother, he is a consummate traveler who blends in anywhere with his one-world look and winning smile. During his twenties, he guided clients up mountains in Alaska, Nepal, South America, even Antarctica. In his thirties, he served as a refugee-camp volunteer in Africa and the Middle East. Now in his forties, he's a physician's assistant in Anchorage. For a few years he visited every continent once a year, including Antarctica. He even flew to the South Pole in winter as a medic for an emergency evacuation, receiving a letter from President Obama for his efforts.

In 2011, another scientist and I were headed to western China to search for Tibetan ice worms on a month-long expedition. Three days before our departure, my collaborator called to say he couldn't go. Minutes after hanging up, I texted Thai: Can you go to China on Tuesday?

Thai texted back a minute later. China? Sure! Let me check at the clinic.

Given leave from work, Thai applied to the Chinese Embassy for a visa. They refused his passport. "Too dirty," they said. He would have to get another one, but felt confident he could, and would catch me in China a week later. We agreed to meet at an airport in remote Yunnan Province.

In Yunnan, he walked off the airplane with a young woman. Their body language and animated conversation said they were old friends. Spotting me, he grinned and hugged hello. "Hey, Roman! We made it!" He turned to the young woman, asking me, "Hey, could we give . . . ," but stopped mid-question to flash his heart-melting smile at her to ask, "What was your name again?" He repeated her name and finished his question, ". . . a ride into town?"

That was Thai: making friends wherever he went, comfortable with whatever was thrown his way. It was also like Thai to drop everything and come down to help me.

BY THE TIME he and I landed in Puerto Jiménez it was Friday afternoon, July 25. The red-eye from Alaska by way of Georgia had left me dimwitted and under-slept. Having Thai along with his outdoor skills, problem-solving abilities, and collaborative nature reassured me. We would find Roman.

We headed to the Iguana Lodge, a few miles beyond Puerto Jiménez. Nestled in a beachside forest, the Iguana hosts its guests in a handful of screened-in cabanas and eclectic structures. The oldest building—the Pearl—is a restaurant and bar, with upstairs rooms and a grassy lawn fronting a palm-lined beach. The newest building is a two-story, open-plan yoga studio with a poolside veranda. Between the pool and the Pearl is the biggest building. This central structure is a postmodern hut with a round thatched roof, a cool, shady downstairs of stone tile, and an upstairs for more formal, open-air dinners. Its office has a landline, computer, and printer we would come to depend on.

Toby and Lauren Cleaver, the American couple who own the Iguana, greeted us. Parents of adult children themselves, they expressed their condolences with sympathy and a sincere desire to help. Both are well respected by their local employees and, like Thai, would be indispensable in my search for Roman. But Thai's help would last only a few weeks. The Cleavers' unwavering support would stretch to months and years. Iguana would serve as our base camp on nearly every trip to the Osa.

Five-foot-two, blond, and smiling, Lauren was fit and fiery and spoke fluent Spanish with a distinctly American accent. Constantly in motion, with a big heart and a sense of justice, she offered to help in any way and, unlike everyone else, she was ideally positioned to do so. Her staff of twenty from the Osa, most of whom spoke English, both liked and respected her, as did former employees who'd moved on to other jobs.

Both Lauren and Toby had been attorneys who needed to escape the ethical ambiguities they faced as defense lawyers in Colorado. They'd bought the Pearl twenty years earlier, renamed it the Iguana Lodge, tripled its size, and added a pool. The Cleavers' sharp, practical knowledge of how Costa Rica functions, together with their extensive network of connections, would be invaluable during our search.

Thai and I drove to the office of the Ministry of Environment and Energy (MINAE), the government agency in charge of Corcovado National Park—a low, one-story gated compound next to the airport. A uniformed man led us to a spare room with a dozen chairs and tables pushed together. A dozen or so men huddled in groups speaking softly. Thin senior MINAE officials in tan uniforms contrasted with husky, young Cruz Roja volunteers in navy vests marked by a red cross. The local police stood by silently in black boots and side arms, their ball caps emblazed with "Fuerza." In Corcovado, where criminals are common, MINAE, Cruz Roja, and Fuerza search as teams.

Someone had put together a poster titled "*Muchacho Perdido*" with "Missing Person" in English just below. Roman grinned in glasses and a scraggly beard. I'd sent the photo only two days before and already it was plastered all over Puerto Jiménez with his name "Cody Roman Dial" and his weight and height in both Spanish and English. The sight of the poster everywhere both reassured and troubled me. Something was being done: people were looking. But Roman was missing and that left me anxious to do something myself. Standing there in the MINAE building was not enough.

By now the police should know where Roman stayed in Puerto Jiménez.

Given the email I'd sent, I expected a debriefing from the Cruz Roja about their search up the Rio Conte, a thirty-minute drive away. Instead, I was questioned by a pudgy, balding, middle-aged man wearing an orange shirt over long sleeves. Like me, his face unshaven, he perspired in the un-air-conditioned room. He introduced himself as Dondee.

I reached out and shook his limp, damp hand. Thai would

translate, as my Spanish was useless. I thanked Dondee for helping. He nodded, eyes closed, and asked, "When was the last time you saw Cody?" It sounded wrong to hear Roman called Cody, a name used by his aunts, grandmothers, and those who only knew him from official documents.

"The last time I saw him was in Mexico, in January. But he emailed me every couple of weeks since then. In his last email he said Corcovado required a guide. But he didn't want a guide. He didn't use them in six months of traveling." I recited the detailed route information he'd sent about the Conte and Rio Claro. "He should have been back ten days ago."

As Thai translated, Dondee pursed his lips as if he didn't believe me—or worse, that he wasn't listening. He responded by asking if there'd been any unusual behavior, as if Roman was just a twenty-something kid who hadn't been in touch with his parents for a while. Hell, I'd gone months without contacting my parents when I'd been his age. But Roman wasn't me.

"Roman always tells us where he is going. Then he tells us when he gets back. This time we have heard nothing about getting back. That's unusual. That's why we're here," I reiterated, annoyed.

Dondee motioned for us to sit. He leaned back, arms folded. Thai translated: "He's asking if Roman does drugs." This took me aback. *Has Roman picked up new habits?*

If Roman was anything like I'd been in my twenties, then he'd tried plenty of drugs. But he had always seemed uninterested in them. As Peggy would say, "He takes good of care of his body. He doesn't want to put drugs into his system." In Anchorage, he lifted regularly at the gym and liked to run. He drank alcohol, sometimes dipped tobacco, and smoked an occasional marijuana or tobacco cigarette, I suspected. But the "dirty hippy" comments in his emails suggested he hadn't started using.

"No, he doesn't do drugs. He drinks. But of course, anything's possible. It would be a big change in character, though."

Dondee went on. "Cody was seen last week walking on a trail to Carate with a well-known drug dealer. He came back to town, paid him at an ATM in Puerto Jiménez, then left to go surfing in Matapalo."

What? This can't be true. Now I was shocked.

Did he make up his trips across El Petén and La Moskitia? Was going-into-Corcovado-without-a-guide a lie? Why hasn't he written us? It's been weeks. Travel changes people, for both good and bad, but how can this be our son?

Dondee's story didn't fit. Roman knew more about tropical ecology at age eleven than most of my college students. He hung out with friends he'd known since kindergarten, packrafted rivers, studied molecular ecology. He hugged his family and friends. To change his character so fundamentally, then lie about it to us all seemed to me not just unlikely, but *fucking* impossible. *Besides, why would he need a guide now after walking across El Petén alone?*

With Thai translating, I tried to explain again that Roman wouldn't have taken a guide on a trail. All his emails had emphasized that popular tourist destinations held no interest in themselves. They were access points for a string of creative, independent adventures across Central America. But showing Dondee emails or explaining Roman's travel style didn't change his mind. The more I tried to persuade Dondee, the more he resisted.

I wanted Dondee to help. He and the others were there *to help*. I was so very grateful to them for that. Still, the Cody they described and the Roman I knew were two very different people. Conventional wisdom holds that parents simply don't know their children well enough to predict their behavior. But with Dondee there was more. He had a self-importance beyond his role as leader of the search. Then, I realized, we had met before.

In 2002, we had both competed in an adventure race in Fiji. He had been on a Costa Rican team that struggled, like most teams, on the first day. I hoped our shared experiences then might create common ground. Instead, it seemed to cast me as a competitor. But this wasn't a race between Dondee and me. We were on the same team in a race to find my son as quickly as possible.

Knowing Roman well, better than anyone, I could help. We had walked on and off jungle trails together since he was three in Puerto Rico. We'd been to tropical Asia, Australia—even to Corcovado twice. It was difficult to articulate the depth of these experiences without sounding both pretentious and arrogant, but my intuition would offer more insight than two dozen Cruz Roja volunteers.

Dondee returned to his computer. A Cruz Roja volunteer sat next to me. "Are you offering a reward?" he asked in clear English.

"No, not yet."

"Good. There was another American, David Gimelfarb, who disappeared five years ago in another national park. He was missing for months and nobody saw anything. Then his parents offered a reward. Suddenly there were sightings everywhere, even in Nicaragua and Panama. But it never led to anything. You see, gringos with blue eyes and blond hair—they all look the same."

The Gimelfarbs' son had gone missing from a simple two-mile trail hike in Rincón de Vieja National Park near the border of Costa Rica and Nicaragua. The Gimelfarbs' $100,000 reward offer caused problems for everybody. Its only outcome was false information about the missing boy and false hope for his parents.

The day dragged on. People came and went. They talked quietly, ignoring me. The success of a search-and-rescue effort comes down to the first few days, the first few hours, often to the initiative and luck of just one person. This I knew from experience.

By the time the sun dropped like a rock at six-thirty, the only things we'd learned were that we shouldn't offer a reward, that Cody was seen with a drug dealer, and that no one had looked on the Conte, the river where Roman said he would start.

"Thai, ask Dondee if they found where Roman stayed in town." Dondee shook his head.

The answer shocked me. After two days of searching, it seemed they should know, yet they didn't. "Thai, let's go," I said. "There's nothing for us here."

CHAPTER 22

The Corners

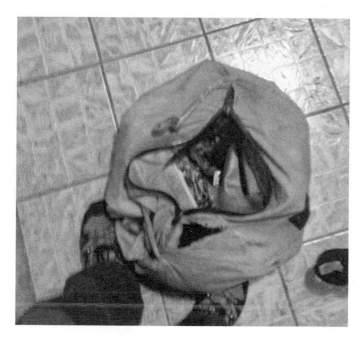

The yellow bag, Corners Hostel, July 25, 2014.

Thai and I left into the night to find where Roman had stayed. We made our way to each of Puerto Jiménez's half-dozen hostels. In fluent Spanish, Thai asked the proprietors if they had seen the young man in the photo we showed. An hour after leaving MINAE headquarters we crossed the only paved street in town, walked past the long-distance bus stop, and arrived at the Corners Hostel. Heavy metal bars enclosed the two-story building up to its tin roof. In front, a picnic table sat beside a small, empty parking lot.

We walked in. An old lady about four and a half feet tall shuffled out in slippers and a simple blue smock patterned in plaid. She was Doña Berta, the owner. She had short-cropped hair, milky blue eyes, and a warm smile, but no English. Thai handed her the photo and asked if she'd seen the young man. "*Sí.* He stayed here, in the dormitory," Doña Berta said in Spanish.

My heart raced. *We found where he stayed! Maybe he's coming back.* Doña Berta showed where Roman had signed in. There, on July 8, he signed his given name *Cody Dial* next to his passport number. This evidence of him comforted me, even if it was just his neat, small-lettered handwriting. I looked at the computers in the office for guest use, wondering if he had typed his emails there. "Ask if the police came by."

"No," Doña Berta responded. Thai and I were the first to ask about him.

"Had he come back?"

"No," she said, "but he left money for his return." She opened a different notebook. Her diminutive hands pointed to an entry in the ledger. He had paid for a dorm bed and was coming back.

"Did he leave anything?" I asked, thinking of all the trips when we'd left things in hotel or airport storage as we headed for the mountains, rivers, and jungles from Australia to Alaska. Doña Berta led us out to a caged-in corner of the building. Immediately I saw the small yellow duffel bag marked "Forrest McCarthy, Jackson, WY." Another wave of warmth and excitement passed over me. The familiarity of his things made him feel close.

Where is he? What is he doing? When will he be back?

Inside the cage was a big backpack, too, but I didn't recognize it. It belonged to another traveler, I surmised, and ignored it. Instead, I pawed through the contents of the yellow bag, looking for answers. Inside was a red spiral notebook. I tore out a page and I wrote him:

Friday 7/25 8:30 PM
   Rome, We were worried when we didn't hear back after 5 days, so Thai and I came down looking for you. Email or go to Corcovado Park Headquarters. There's a big search on for you. Hope you are OK!
Dad

Back at MINAE headquarters, we got in our rental jeep and drove to the Iguana. Toby and Lauren were waiting, eager for an update. We told them how we'd found Roman's hostel, that the little old lady there had said he'd planned to return but never did, then mentioned to them the story of the drug dealer and hike to Carate.

As locals, they knew of the guy and his name. "We'd heard that, too," Toby said, "that Cody was seen with Pata Lora. Our breakfast cook has a relative in Piedras Blancas who saw Pata Lora with your son."

"Pata Lora?" I repeated.

"Yeah, Pata Lora," he said. "Short for *Pata de Lora* or "Parrot Foot," a reference to his limp. He's a really bad guy. A thief. Into drugs. He comes from a big family on the Osa that's done well. But he's the black sheep. Nobody likes him. You can't trust him. Even his own parents disowned him."

Lauren interjected. "It's so like the Fuerza not to even find the hostel where your son stayed. The authorities never really do an investigation quickly or effectively. When our friend Kimberly was murdered a few years ago at her house, they never even found a suspect. We had to hire a private investigator to find out who did it."

After dinner Thai and I went to our room. It had been a long day. In the darkness and heat, I tossed and turned below a wobbling ceiling fan, trying to reconcile Roman's last six months of emails with the story of Cody hiring a drug dealer as a guide outside the park.

THE NEXT DAY I told Dondee that we had found Roman's things at the Corners Hostel. He didn't care. The prevailing narrative on the Osa that Dondee now pursued involved Pata Lora, the same well-known twenty-seven-year-old thief, sometime miner, bootleg guide, and general Osa ne'er-do-well that Toby had told us about.

In this story, Cody and Pata Lora walked a horse trail from Dos Brazos, through the off-road mining community of Piedras Blancas, then onward to Carate by footpath, a journey entirely *outside* of Corcovado National Park. From there, they returned by colectivo to Puerto Jiménez, where Cody paid for Pata Lora's guide services from an ATM there. Then, Pata Lora said, Cody went surfing at Matapalo, the rocky cape with the best break on the Osa. These were details offered by Pata Lora himself to Costa Rica's version of the FBI—Organismo de Investigación Judicial, or OIJ, pronounced "oh-ee-hota"—who had questioned him about his travels with the gringo. According to Dondee, a nature guide named Roger Muñoz saw Cody with Pata Lora coming out of the jungle in Carate.

This news was good, if a bit perplexing. Cody Roman was around, just acting strangely. Since Mexico, he'd been quick to answer our emails. But after Peggy had responded to Roman's email on July 9, calling him the next Thai Verzone, he hadn't written back. *Roman wrote us the day he exited El Petén and La Moskitia. Why hasn't he emailed us now?*

Thai and I headed to Carate at ten, following Dondee in his Cruz Roja Land Cruiser to meet Roger Muñoz. Driving through Puerto Jiménez, I did a double take every time I saw a young man in short hair and glasses who wore flip-flops and a tank top. They all looked—even to me—like Roman. "The Cruz Roja volunteer was right," I told Thai. "Every gringo looks the same."

The potholed road to Carate punched abruptly from Puerto Jiménez's residences into cattle land. To the east, a gentle surf sparkled in the sun. Ahead, the road passed through a tunnel of fig trees for miles, their fat trunks strung with barbed wire as living fence posts. We bumped along as fast as the corrugated dirt road allowed.

"Thai, thanks for coming down. I wouldn't have found that hostel without you."

"Sure, Roman. I'm happy to help. I just hope we find Roman soon."

"You know, I'm glad everybody seems to think he's around, but it seems weird that he hasn't contacted us, or even gone back to get his things at the hostel."

Thai nodded.

"And if he was headed out surfing, why would he leave all his beach stuff in Forrest's duffel at the hostel?" It seemed unlikely he'd leave sunscreen and his dive card behind.

"Yeah, and it's weird he'd hook up with this guy Pata Lora. That doesn't sound like Roman to me."

I *wanted* to believe the Pata Lora story, but it just didn't fit.

Science confronts hypotheses with evidence through the process of disproval and I was more than willing to be proved wrong, especially if it meant Roman was okay.

"Thai, I don't know what's going on, but I sure hope these people are right about Roman."

CHAPTER 23

Carate

Tamandua, along the road to Carate, July 26, 2014.

Carate is little more than a string of beach camps, locals' houses, and expensive second homes hidden from view. Parallel to the palm-lined beach, there's a long paved airstrip where clients arrive for upscale vacations at rustic eco-lodges. Mountains rise steeply above the pounding surf on the beach. Scarlet macaws cruise over the trees. A popular jumping-off point for day hikes into Corcovado National Park, Carate's colectivo stop is at the end of the road, next to a pulperia that sells drinks and snacks for those waiting to catch a ride back to Puerto Jiménez.

We parked and headed down the beachside path leading to the La Leona Ranger Station at the Corcovado boundary. Slender Roger Muñoz met us on the trail. In his twenties, his broad smile was open and his short hair and wide ears gave him the clean-cut looks of a guide who likely earned good tips.

"No, I don't think it was him," Muñoz offered in clear English, looking me—the father—up and down while inspecting a recent photo of Roman. "He wasn't so tall. I wouldn't have noticed him if he wasn't with that guy, Pata Lora. Pata Lora is a bad guy, not a real guide."

Still, Carate *was* where Roman intended to end his hike. And Roger *had* seen a gringo with Pata Lora on July 15, when Roman would've reached the road, and the day Roger signed La Leona's logbook. The time and place fit. Just not the gringo.

After saying good-bye to Roger, we encountered a group of high school kids from the U.K. They had walked over from Piedras Blancas, the off-road/off-grid mining community halfway between Dos Brazos and Carate. They were waiting for the big Mac truck that operated as the colectivo. We asked if they'd seen any other gringo hikers. No, they said.

A local guy in shorts and a T-shirt, wearing rubber boots and smelling of alcohol, hoisted himself into the front seat of the colectivo after it arrived. In a conspiratorial voice, he leaned out the window and told Thai that the kid we were looking for had been seen with a very bad guy on the trail from Piedras Blancas. Then the truck pulled out and drove back to town, carrying the British kids, the drunk miner, and the story to Puerto Jiménez.

FROM 2009 TO 2011 four ex-pats were murdered on the Osa. The two Austrians in their mid-sixties had been living and buying gold in Dos Brazos when they went missing during Christmas 2009 from their blood-spattered house. Their vehicle was gone, too. Two years later, a flooding stream washed bones out of the beach where the murderer had buried their dismembered corpses. The same year, the fifty-three-year-old Canadian friend of Lauren and Toby named Kimberly Blackwell was found beaten and shot at the gate to her home and cocoa farm between the Barrigones and Conte Rivers, near where Roman had said he'd start his hike. Later that year, fifty-two-year-old Lisa Artz, an American and another friend of the Cleavers, was suffocated when thieves stole her laptop and iPod.

While these murders ultimately resulted in convictions, such justice was rare. In fact, it took a private investigator hired by Lauren, Toby, and other friends of Kimberly Blackwell to identify the killer. Overall, statistics show that less than 5 percent of murder charges in Costa Rica end in conviction: nine times out of ten, perhaps, people get away with murder. The Osa works hard to mask this darker side, offering surfing lessons, yoga retreats, and guided walks. Still, some locals often rely on illegal activity for their livelihoods and the people who know the jungle best include poachers and gold miners who somehow avoid the poisonous snakes, tree fall, mudslides, wild animals, and flash floods while dodging park rangers who burn their illegal camps.

Roman might well have overlapped with characters connected to the Osa murders. An early suspect in Kimberly's murder lived in the foothills above the Rio Conte. Pata Lora's cousin was sentenced to fifty years in prison for killing the two Austrians. Cody was reportedly seen near Matapalo where Lisa Artz was murdered in her own bed.

While we were in Carate, Doña Berta, the little old lady at the

Corners Hostel, had changed her story. She now remembered Cody *had* returned, then left again. We accompanied Dondee and Tony, a Cruz Roja employee stationed in Puerto Jiménez, back to the Corners to investigate. The pair studied Doña Berta's cryptic entries, asking questions that to me sounded like interrogation.

I'd hoped to see something concrete in the ledger: "Dial" or "Cody." Instead I saw "XXXX" and "₡5,000 pago" marked in green highlighter. Fingers pointed at text, flipped pages, and settled on "Martes 22 Julio." The three concluded that Cody had returned on Monday, July 21, then left again for Dos Brazos on Tuesday, July 22, leaving money for a bed on his return on Wednesday, July 23. Today was Saturday, July 26. According to this account, Cody had been here only four days ago and was expected back any day.

The relief of this news settled over me like a warm blanket on a shock victim. I smiled broadly at Thai. I looked forward to seeing my son. It had been six months since our rafting trip, the second longest gap in his life without physical contact between us, without a hug, a shared meal, a pun, or a grinning story. I was sure he had some new tales to tell from his weeks on the Osa.

Prior to our arrival, Cody sightings had come from all over the Osa Peninsula. Cody had been seen wearing a safari outfit between Carate and Matapalo. A bus driver had dropped him off at Dos Brazos; miners had seen him in Piedras Blancas. Pata Lora had claimed he was with Cody in Puerto Jiménez and that Cody went surfing afterward, where he was seen near a Matapalo bar.

Listening to all these sightings, it seemed easier for all to disregard Cody as an irresponsible twenty-something who was too cheap to hire a real guide than to accept Roman as lost or injured in Corcovado's wilderness. The sentiment was "Let's just wait for Cody to show up. If he doesn't, then he doesn't want to."

Besides, the kid sounded foolish: illegally in the park, alone and off-trail in its wilderness. To look for Roman in a trackless jungle of poisonous snakes, lawless miners, and few trails was like looking for a needle in a burning haystack. The more I claimed that the Roman I knew differed from the Cody that everyone cast as a stereotypical gringo kid, the more they pitied me as a father in denial, the first stage of grief.

This reaction reminded me of an incident a decade ago. One of my former students at APU, a popular, easygoing kid named Joe, had taken up climbing. Joe and a more experienced mountaineer were climbing a local Anchorage peak unroped when a cornice collapsed and sent Joe thousands of feet to his death. When his father heard of the accident, he rushed to Alaska, arriving at the airport ready to head onto the glacier where Joe had fallen, thinking he might still be alive.

The father had no experience with glacier travel. He brought downhill skis and boots unsuited for skiing uphill. Although he could, of course, conceive of the danger and knew he lacked the skills of professionals who had been unable to find his son, he was still a father who loved his son deeply. His instincts had implored him to act. The father never set foot in the mountains, perhaps talked down by the leader of the search, or simply aware of his own limitations. After Joe's father returned home, I telephoned him, partly to give my condolences, but mostly to empathize father to father. "I have a son," I said. "I can't think of anything worse than losing him."

Eventually, I gave up kicking off Cruz Roja's warm blanket.

Giving in kept my shock at bay. It felt good. Cody was everywhere and doing fine. He just wasn't contacting his friends or parents.

Relieved, I emailed Roman that night:

> Looking for you. Everyone is. Wished you'd emailed us when you
> had the chance. They say you were with this guy Pato de Lorra
> or something like that. And he has been arrested and is being
> questioned. They say you've crossed the mountains twice now. With
> Pato de Lorra. Practice for Darien? Hope to see you safe and sound
> and soon.

I had Thai ask Tony, who lived in Puerto Jiménez, where was the best place in town to eat. Feeling gracious, I wanted to treat Dondee, Tony, and Thai to a meal. As we waited on our dinners of seafood, rice, and plantains, a gentle sea breeze carried the night air into the restaurant. I stepped out to the busy waterfront and called Peggy on my cell.

Waiting for the call to go through, I gazed out into the tropical night, looking hopefully at every young man in a tank top and short hair, expecting him to say, "Dad! What are you doing here?"

Peggy picked up. Just hearing her voice soothed me.

"It looks like people have seen Roman around. He left some stuff at the hostel and now the old lady who runs it remembers him coming back. It looks like everything's okay. I hope he's not mad that I'm here." We both chuckled.

"Oh, good," she cooed. "It sounds like he's back, but just not checking in."

"I still think coming down was the right thing to do."

"Of course, it was. He wouldn't have written us if he didn't think we'd come down to help. You had to go down to be sure he was okay. You're his dad. What are you going to say when you see him?"

"I don't know. Maybe, *Let's do a hike in Corcovado?*"

We both went quiet. Our silence admitted that the whole thing was unlikely. But it felt good not to worry. To think he was safe. That he was okay.

CHAPTER 24

The Helicopter

Corcovado mountains from the air, July 27, 2014.

After dinner, Dondee informed us there'd be a helicopter search in the morning. This suggested in actions, if not words, that somebody besides me was not fully convinced by Pata Lora's story. Or that Doña Berta from the Corners Hostel had convinced Dondee that Cody had gone back into the jungle a second time—and not come out.

More likely, the helicopter search was a response from American pressure to look more thoroughly. Since the day I left Alaska, the state's lieutenant governor, a friend of mine named Mead Treadwell, had been pushing for U.S. National Guard involvement. Ultimately, Mead's push for military assistance would reach the four-star general in charge of the Southern Command, General John F. Kelly, one step below President Obama's Secretary of Defense.

Thai and I went back to the Iguana and heard the story of Pata Lora there, too, a narrative ossifying like a plaster cast across the Osa. Whenever one local whispered "Pata Lora," another nodded solemnly, or twisted his face in a grimace. Maybe frontier justice was a simple matter of picking the local pariah for the most recent crime. Innocent or not, truth be damned, at least they'd be rid of the rat.

Depending on who was talking, Pata Lora was currently being held for drugs, theft, even murder. Pata Lora told the OIJ that the gringo paid him from an ATM on July 16, after Roger Muñoz had seen them in Carate. Here was something tangible that we could check: when did Roman last withdraw money?

I filled my notebook haphazardly with scribbled names, numbers, notes, and quotes. Still half convinced Cody Roman was around somewhere, but just too stoned to check his emails, I'd written in the margin, "We could just wait for him to walk out—but too many people involved, too much momentum."

Back at the Iguana, Thai and I shared a room stuffed with gear. He wore his long black hair in a ponytail that reached past his shoulders and a choker of dzi and other beads he'd bargained hard for on our trip to Tibet the year before, when we had searched Himalayan glaciers a second time for ice worms. Thai was an adventurer who could do everything—climb, boat, ski, paraglide, hike fast, mountain-bike, navigate, save lives. Importantly, with his golden-colored skin and almond-shaped eyes, he fit in everywhere he went. People welcomed his warm smile, easy laugh, and honest enthusiasm.

"We gotta get to the Conte River, where Roman said he was going," I implored. "Cruz Roja doesn't seem to believe he ever went there. And this Pata Lora story—much as I wish it were true—it doesn't seem right. If Roman came out, he would've contacted at *least* his friends, if not me and Peggy."

Thai frowned sympathetically. "Yeah, I know. But first we have to establish credibility with Dondee and the park service. We have to show we are capable in the jungle without making mistakes. Accidents happen and with us wandering around out there . . . well, we could create another rescue situation. That's their concern."

Thai was right, of course, but my son was missing. My son, who I hoped would be twice the father I tried to be. I'd seen Roman's tender side, an easy patience, and playfulness with little kids—qualities that are rare in young men. Besides, parents who want kids—like Peggy and me—usually want grandkids one day, too.

Instead of wringing my hands in town, where all I could do was mull over why this Pata Lora story wouldn't go away, I wanted to search in the park myself. That's why I had come, even if everyone saw me like Joe's dad at the Anchorage airport carrying downhill skis, desperate to find his son on a wild mountain glacier.

By our second morning in Costa Rica, the euphoria that Roman would walk out at any moment was gone. The previous night's relief on the phone with Peggy was a momentary peak on a roller coaster now plunging toward fear. Thai and I drove to the search headquarters where we were briefed— and questioned—by the Cruz Roja, MINAE, and Fuerza.

Cody hadn't appeared. With posters everywhere, it seemed impossible for him not to have seen his own face somewhere on the Osa.

Dondee stood in the briefing room full of searchers and told us all a helicopter with infrared search capabilities was winging its way from San José. He moved among the Cruz Roja volunteers, nodding his head when spoken to, hands on his hips or shaking a finger when speaking, his arms giving orders. Thai and I noticed we weren't the only ones uncomfortable with Dondee. "He's like a ballerina needing attention," a local Tico nature guide whispered.

The Cruz Roja were mostly pale men from San José. They sat in chairs wearing earnest expressions and khaki uniforms tucked beneath their belts. Along the wall stood lean, scrappy guys, dark from the sun, shirttails free. These were the local rangers who knew the park, its miner and poacher trails, its ridgetops and canyon bottoms. They were the ones I wanted—especially the young ones—for a search team of my own.

The Fuerza asked me if we had any enemies. Was there anyone who'd do us any harm? I thought of emails Roman's friends shared that mentioned, finding a Costa Rican girl. I wondered if he'd pissed off a boyfriend or even a husband somewhere. Roman also had a dark, aggressive side I'd always suspected but never seen. Some bravado in emails to friends hinted at it, and now my imagination ran with the suggestion.

"No," I replied. "None that I know of."

"Did he have a phone?"

"No, it was stolen in Mexico."

"A GPS?"

I'd heard the gringo with Pata Lora had a GPS. "Not unless he bought one down here."

"How about Facebook or social media?"

After the loss of his girlfriend and Vince's death, Roman stopped using Facebook. "It's just a bunch of people who don't do anything, bragging about it when they do," he explained. I had wondered, though, if it wasn't because of his ex-girlfriend's social media presence.

"He keeps in touch with email," I told the Fuerza.

Again, I brought up the emails he had sent about the guide requirement, his intentions to go solo and off-trail, that maybe he was preparing to cross Panama's Darién Gap.

"The volunteers are excited to find your son," Dondee said, looking me in the eye. He continued in English. "These guys live for this. *They live for this!*"

"*That's good.*" I nodded sincerely, smiling. "*That's what we need.* I'm so grateful for all these people who are helping. I don't want them to stop."

Then I reemphasized that Roman said he was heading in off-trail.

"The park superintendent says that off-trail travel in Corcovado is 'illogical,'" Thai translated. The whole thing felt like an exam and I was giving only wrong answers.

"Tell them, 'When he traveled off-trail in Guatemala he marked his route.' "

"How did he mark his route?" a ranger asked.

"By cutting with a machete," I said, though it felt like another wrong answer.

"Everybody makes those marks," Thai translated back, continuing: "Did he ever leave little reflectors?"

"What?"

Thai spoke in Spanish with a ranger. "Like thumbtacks. Did he ever leave thumbtacks? The guy with Pata Lora had left reflective thumbtacks in trees."

This actually sounds like something Roman might try. It seemed clever, like the way he would mark small mammals by clipping their fur during live trapping projects. Reflective thumbtacks would be simpler than a machete but much harder to follow, except in the dark with a headlamp, perhaps.

"It's possible," I said, cautiously. *But why would he mark a tourist trail while carrying a GPS and following a guide? It's like wearing a belt with suspenders on a one-piece suit.*

My mind sorted through these new questions as I left headquarters and headed for the airfield where the helicopter waited. The sturdy, compact black machine held five of us. No one but me spoke English. The effort reminded me of another search a decade before.

In Alaska, where the woods are open, I had once chartered a helicopter to search for three teams overdue in a Wilderness Classic I'd organized. One racer had positioned herself on a bald hill with a smoky signal fire; she was easy to find. I never found the others hidden by Alaska's short-statured forest. *How will we find someone beneath a forest three times as tall and five times as thick?*

WE LIFTED OFF. I looked down at a rural landscape of cattle pastures and subsistence farms. The patchwork gave way to steep, forested mountains. We buzzed low outside the park between Dos Brazos and Piedras Blancas, where Cody reportedly followed Pata Lora. The only surfaces visible were the flash of streams, the dirt of landslides, the pastures of Piedras Blancas. Mostly, I saw billowing green crowns and the occasional slender white trunk in a tangled forest canopy. Banana-like heliconia plants obscured old landslides. Bracken ferns covered new ones.

The most recent slides, still raw with bare dirt, drew my attention. *Such a dynamic landscape.*

We strained to see some sign of Roman. A flash of color that he might position on a creek's gravel bar or still-water pool, or some gear he might drag onto the fresh surface of a new landslide. But there was nobody. Not a miner, not a tourist, and no lost and injured young man.

I couldn't help but appreciate the effort, the cost, the futility. I wanted to see him there. But peering at the surface of an ocean of green foliage made it clear why everyone hoped Cody was on the trail to Carate, smoking pot with Pata Lora, or hanging out in a bar in Matapalo. Finding him *outside* Corcovado at least seemed possible.

Fourteen minutes after leaving Puerto Jiménez, we reached the Pacific. We circled over its aquamarine near-shore waters, then dropped off a Cruz Roja volunteer on the beach near Carate. It felt useful to have so many eyes on the ground. We flew back to Puerto Jiménez over the park's north and west side. From the air, Corcovado's mountain range looked like a clenched fist, palm down, knuckles as highpoints, the back of the hand sloping toward Dos Brazos, the finger joints forming cliffy canyon rims above the north and western rivers: Sirena, Rincon, and Claro—Roman's destination. Here and there enormous trees with yellow blossoms punched through the canopy and waterfalls plunged off the summit escarpment. It was beautiful, futile, heartbreaking. *He's likely out of food by now.*

I hoped that the helicopter signaled to him we were coming, that we would find him. The helicopter flight was useful in another way, too. It confirmed that to find Roman I needed to follow his trail into the jungle and retrace his route. Fifty minutes later we landed back in Puerto Jiménez. Officials would ultimately log over nine generous hours of helicopter time. I was grateful for the effort, even if disappointed by the not unexpected outcome.

After the flight, Thai and I returned to the Corners Hostel to see if Roman had come to pick up his things without telling us. Doña Berta said she was sorry that my son was missing. She said God would help. Then she asked me to take the yellow bag and not come back.

CHAPTER 25

Rio Conte

Pancho in Rio Conte Valley, July 30, 2014.

I spent the rest of the day in the car on my cell phone talking to the embassy, the FBI, and Peggy. I hated talking on my cell. It came to signify empty promises, disappointments, expensive bills. The FBI relayed that the last email Roman sent from his email account was on July 10, the day after he'd written us. He'd written his college sweetheart two words that morning: so weird. Peggy told me that she and Lieutenant Governor Tread-well had managed to learn from reluctant bank employees that Roman made no financial transactions after July 9. If Pata Lora

had been paid at a Puerto Jimenez ATM on July 16, it wasn't
by Roman.

Roman's Costa Rican case was officially a missing persons
one, so the FBI couldn't get involved in the investigation.
There needed to be evidence of a murder, a kidnapping, or
extortion. The embassy would work as an intermediary among
us, the FBI, and the OIJ, but our son was in the hands of the
Costa Ricans. Jorge Jimenez—at twenty-seven, the same age
as Roman and Pata Lora—was the OIJ investigator in charge.
He'd come to the Osa from Ciudad Neily, a two-hour drive
away, for another case.

Lauren had a knack for tracking people down while keeping
the Iguana running smoothly. She called the OIJ agent and arranged
a meeting. Jorge Jimenez met Lauren and me at a pulperia along
the Osa highway. A baby-faced young man stepped out of a dark
SUV. He wore nice black shoes, slacks, and a button-up shirt that
was open at the collar. Jorge was polite enough, but tight-lipped,
too. He said Roman's was a missing persons case, with no evidence
of foul play. There was no body, for instance.

"What about Pata Lora?" I asked. "I hear you have him in
custody."

"Pata Lora is being held on another case, unrelated to this.
We've heard the stories and questioned him."

"Was Roman with him?"

"Pata Lora," he offered, "is not a reliable witness."

Jimenez asked about Cody's credit cards. There were none—
just Roman's ATM, library, PADI dive, and ID cards left in the
yellow bag. The detective asked me about the red notebook that
had also been at the Corners Hostel with its entries after July 10.
Jorge did not know these entries were from 2011, when Roman was
in arctic Alaska working a field job. They concerned snowy owls,
trapping shrews, baiting isopods, and working with colleagues.

He'd also written poignantly about heartbreak after his girlfriend had left him the year before.

The OIJ's interest in Roman's three-year-old journaling was another example of an uninformed lead, like Cody hiring Pata Lora to hike a trail outside Corcovado. I appreciated the effort, but the view from the helicopter had presented a sobering reminder that finding Roman in the tangled landscape would most likely take skill and luck from a small team piecing together his route. Roman had declared his intentions. Why wasn't anyone paying attention to them?

Dondee had prevented my access to the park, but influential people intervened. Between Lauren's calls to local officials and a letter of introduction from Lieutenant Governor Treadwell to a well-connected Costa Rican named Juan Edgar Picado, it looked like we were in. Lauren hand-picked a ranger and guide whom I would pay to accompany Thai and me to the Conte.

The authorities nixed that idea. It came too close to offering a reward. So Dondee himself drove Thai, me, and Lauren's recommended ranger, Pancho, up a steep, bumpy dirt road to the La Tarde field station on the north side of the park, close to the Conte. La Tarde, I understood, would provide a base from which to search the Rio Conte.

Early the next morning, we left La Tarde and walked half a mile to a trail junction. One fork followed the central ridge, called Fila Matajambre, to Dos Brazos. The other fork dropped into the upper Rio Conte. Dondee said he'd wait there with Pancho's pack while Pancho led us into the Conte valley. The owner of a small ranch said no gringo had been through in months. I insisted we go to the creek itself, to look and intuit where Roman might go.

We followed the Conte upstream into its headwater canyons. Clear water spilled over bedrock slick with algae and moss.

Philodendrons smeared themselves on near-vertical walls between ferns and small trees that somehow clung to crevices and
nubbins. It was green, hot, and humid.

Pancho urged me to lead, gesturing that I search for sign. I
picked my way slowly, looking for broken sticks, crushed and
broken plants, a foot pad, praying for a candy bar wrapper—
anything that said, "*YES!* Roman was *here*!" But instead of signs
that a human had passed three weeks ago, I found only evidence
that a tapir—the five-hundred-pound, three-toed animal that
looks like a chimera of elephant, rhino, and pig—had pushed
past palms and pepper plants three hours earlier.

We returned to the trail junction. Dondee and Pancho's pack
were gone. Pancho marched back to La Tarde, finding that
Dondee had driven off with our only vehicle. Pancho returned
to us and, irate, declared Dondee *loco*. The three of us walked to
La Tarde and waited.

Dondee drove up hours later. When confronted, he complained
that our two-hour search in the Conte had been too long. He
turned to me. "*You* have to go," he said. "Go back to town and
stay there. You're unstable. And if you try to go into the park,
you'll be arrested."

"But this is where Roman said he'd come," I pleaded. "We
need to look here, follow the Matajambre trail. That's where the
Conte trail leads."

Thai entered a longer discussion with Dondee. "They seem
like they're trying to isolate you, Roman. Keep you in a little
bubble. They say it's because they are concerned for your safety."
Thai had negotiated a search along the Matajambre trail. "You
can't go, but they'll let me go with Pancho and another ranger,
Kique." Thai would spend the night at La Tarde and walk the
Fila Matajambre to Dos Brazos the next day.

A CRUZ ROJA driver took me on the long, lonely ride back to the Iguana, arriving in an afternoon downpour. Alone in my room, I imagined all the ways to suffer and die in the jungle. Quickly, by the deadly bite of a bushmaster or fer-de-lance on the leg. More slowly and painfully by a small arboreal eyelash viper's bite to the face. Crushed by tree fall. Through sepsis in a leg or arm, broken when a rotten log gives way. By dengue, malaria, or other fevers. Debilitated and starved after a fall from a cliff or waterfall. Finally, there are the other *snakes*—the two-legged kind.

In my room I felt sick, with no appetite. I pictured Roman huddled in a soaking wet tent, too injured to travel, eating lizards and bugs. He'd eaten a lizard on his Petén trip. He knew discomfort. He had a level head. He could hang on. I just needed to find him before it was too late. *What have I done?*

Alone and without a task at hand, my eyes teared up and I sobbed, thinking of our family trips to the tropics. It was impossible not to reminisce. Those experiences made up our family lore, our history: hearing gibbons whoop at dawn, handling a flying lizard, eating exotic fruits. *I took my eight-year-old son to Borneo's wilderness. Was that negligence?* It hadn't seemed so then, but now I felt a sharp stab of regret. Not because we had risked his life or Jazz's in Borneo, Australia, or any other place where humans have lived for millennia—Peggy would never have endangered her children's lives—but because of the life it inspired.

My regret was that I had introduced Roman to adventure and the excitement of the wild. Maybe we should have limited ourselves as parents to team sports, Chuck E. Cheeses, the local cineplex. But that would have been impossible for Peggy and me. "What, take the safe but boring route?" she would ask.

"Birth is the leading cause of death," my friend Brad Meiklejohn likes to point out. Still, the cliché "At least he died doing what he loved" is wrong, Brad says. "I most admire those who have done what they love their whole lives and died peacefully in bed at a ripe old age."

The guilt of responsibility is persistent, pernicious—perhaps simply instinctive—in parents whose children are injured, lost, or killed, even after we rationalize or realize that it wasn't our fault at all. For me in Costa Rica, sleep offered the only respite from worry and pain. But when I awoke, before my eyes even opened, the fact exploded as my first conscious thought: *Roman is missing!*

THE NEXT DAY, July 30, a Cruz Roja Land Cruiser picked me up at the Iguana and delivered me to the El Tigre ranger station in Dos Brazos. Eliecer Arce, the head of Corcovado National Park, and Carlos Herrera, the head of the Cruz Roja, were there, too. Thai, Pancho, and Kique would finish the Fila Matajambre trail in Dos Brazos, where we would all rendezvous.

Pata Lora and Cody had started their hike to Carate in Dos Brazos: witnesses had seen the pair there. Waiting for Thai at the end of the road in a small cantina, I could make out the words *Pata Lora* and the Costa Rican word for marijuana, *mota*, in conversations around me. I shook my head. The only way to prove that story wrong would be to find Roman's body. I prayed that wouldn't happen.

The owner of the cantina was named Elmer. He was one of the witnesses and spoke good English. Holding his young three-year-old in his arms, he recounted how he'd seen Pata Lora and Cody walk past his cantina on the trail to Piedras Blancas.

As I was listening to Elmer, Thai ran up. He was smiling, out of breath, muddy. It was good to see him. "Roman, hey!" He reached out his hand, gave me a sweaty embrace. He was visibly excited. "It sounds like somebody saw Roman!"

"What? Where?"

"Down here, just a few houses! A guy met someone in the jungle who said his name was *Roman!*"

The news buoyed me in a way I'd not felt since arriving in Costa Rica five days ago. Here, at last, was evidence.

CHAPTER 26

Jenkins

Jenkins, Dos Brazos, July 30, 2014.

To a Costa Rican, the name "Cody Roman Dial" has both "Roman" and "Dial" as surnames—like two last names. Anybody explaining that they'd *met* our son *Cody* was either lying or mistaken. He'd been introducing himself as *Roman* since the start of our walk across Umnak twenty years before. No Costa Rican would think to call him "Roman" unless they had actually met him. This was another reason Pata Lora's story didn't add up.

Thai led me to a small wood-frame house painted yellow on a concrete pad with a sheet-metal roof. Two little girls peered out the windows. Wearing only shorts, a young man stood erect and muscular with a mustache and a small soul patch under his lip. Like Thai, he smiled easily and broadly. Like Roman, he was twenty-seven. He spoke remarkably good English. I didn't need Thai to translate.

Breathless and excited, I blurted, "Hi! My name is Roman and my friend Thai says that you may have seen my son in the jungle!"

"Yes, that is true." He answered in slow, measured English. "My name is Jenkins Rodriguez and I am a gold miner." He held out his hand and we shook.

"Where did you see him?"

"We saw him in the mountains on a small trail. We have never seen any gringos or foreigners there, so it was very surprising to us."

"When was this?"

"Maybe about fifteen to seventeen days ago. He said he'd been in the forest for two days."

I did the arithmetic. It would have been around mid-July when Jenkins had seen him. But people here didn't keep day planners, diaries, or even time, really. Few wore watches in the villages or jungle.

"Where did you see him?" My hand shook as I scribbled in my notebook.

"On a small creek called Zeledón. It is about three hours upriver from Dos Brazos."

"Did he tell you his name, where he was from?"

"Yes. He said his name was Roman. He was from Alaska and he was a biologist."

I took a deep breath, shocked. Other than Doña Berta at the hostel, this was the first person who I believed when they said they had seen Roman. I racked my brain for what the newspapers had printed, wondering how much of Jenkins's story could have been pulled from news accounts. Certainly, the fact my son was from Alaska, maybe the part about going by "Roman," but definitely not the part about being a biologist. To me, Jenkins's rough dates felt like truth. He sounded honest, with nothing to gain from lying and little to hide.

"What was he doing and what did he have with him? Did you talk to him?"

"Yes, we spoke. The moment we walked up to him he was sitting there, cooking his breakfast on a stove. I think it was some rice. At first, we just walked by, surprised to see him there, and said hello. Then we turned around and came back to talk with him as it was very strange to see a gringo in this part of the forest. He spoke Spanish slowly and was easy to understand, but we switched to English."

At this point I thought of the blue Jetboil stove I'd given him for Christmas. It wasn't in the yellow bag. "What did the stove look like?"

"It was not the kind from around here." Jenkins motioned with his hands that it was taller than wide, and compact. *Like a Jetboil.*

"Who else was with you?"

"At the time we saw him, I was with Luiz and Arley, but we waited for another miner named Coco, who came very soon. There were four of us."

My head flooded with questions. "Did he have a camp set up?"

"No. He said he had come up the river the day before, ran into a waterfall, then climbed a cliff and made camp on the ridge. In the morning, he came down to the little creek to make breakfast. He had a green-colored pack with maybe a rolled-up pad for sleeping. I heard about the missing guy and I thought it was the same one who walked through town. The one Elmer saw. But now I don't think those two are the same guy."

"Did he have on glasses?" asked Thai.

"I don't remember glasses. He was shaved, and a little bit serious. He said he was a biologist and just looking at the many different trees and the different things in the jungle. He asked if he was in Golfito, so I thought that maybe he was little bit confused. He showed me his map. It was about the size of your notebook papers. I didn't want to scare him by staring and making him uncomfortable. I was just friendly with him that day."

My head was spinning. Even the "Golfito" reference made sense, because the "ESRI world topo link" I'd sent Roman labeled the border between the two cantons Osa and Golfito inside Corcovado's boundary. Roman would have been referring to the canton of Golfito on the map, not Golfito the town across the bay from Puerto Jiménez.

I wanted to talk to Arley, Coco, and Luiz. I wanted Jenkins to sketch a map of where this all took place. And most of all, I wanted to go to this Zeledón Creek, Roman's point last seen.

That was where I'd go look, and soon. It didn't matter that Dondee had threatened to have me arrested if I was caught in the park. Nobody could stop me from looking there now.

CHAPTER 27

Zeledón

Vargas and Jefe, point last seen, July 31, 2014.

Jenkins offered to take Thai and me to the point on Zeledón where he had seen Roman. It felt like we were finally on Roman's trail, not just milling around town or sitting quietly in the sparse headquarters at MINAE. If searching the jungle meant going in with illegal miners and other Osa criminals, so be it. Lauren connected us with another man—named Vargas—who was neither ranger nor guide but intimate with the park in ways no one else could be.

Thai, Lauren, and I met Vargas by the bank in Puerto Jiménez. It was almost eight in the morning and the sun's heat would soon force its daily discomfort into everyone's life. A few locals parked their trucks to sell rambutan in the shade of broad-crowned trees that overhung the street.

"That's him," Lauren said as we walked toward the corner. "He was a suspect in Kimberly's death. He's a poacher as well as a farmer. His brother was killed by a bushmaster when they were young. Vargas probably knows Corcovado's mountains better than anyone alive."

Dark from six decades in the tropical sun, Vargas was short and compact with a toothy grin and a mop of black hair. His hand, muscular from a life of farming, grasped mine firmly. He looked me square in the eye. Vargas was in town for business and dressed sharply in a pearl-snap shirt with crisp blue jeans pulled over pointy-toed cowboy boots. He'd taken the bus from his small oil palm farm just south of the Rio Conte.

Lauren told Vargas about the four miners who met Roman in the jungle two weeks earlier. She said the Cruz Roja and MINAE had banned Thai and me from the park, but that we were headed to Dos Brazos and up El Tigre anyway. Vargas shook his head and kicked at street dust with his boots, scoffing at the MINAE and their iron-fisted control of the park.

"Lauren," I said, "ask Vargas if he can go today with us and the guy we met in Dos Brazos who saw Roman up the Tigre. Tell him it's about three or four hours upstream." Lauren rattled off my request in her thick American accent.

"Si." Vargas nodded, looking me in the eye again, but he would need to do some shopping first.

VARGAS AND HIS eighteen-year-old son, Jefe, joined us in our rental Suzuki as we drove the bumpy road to Dos Brazos. It was nine when we picked up Jenkins.

Jenkins lived on the Piedras Blancas arm of the Rio Tigre in Dos Brazos but would lead us up the El Tigre arm into Corcovado where he had his mine and *rancho*, one of the small tent camps of black plastic tarps that were routinely burned by rangers and remade by miners. We parked at the end of the road. A narrow trail above the creek led past a tin shack surrounded by barbed wire where a bicycle leaned against a fence post. A toothless, shirtless guy in shorts called out *Hola* as we walked by.

"That's Pata Lora's uncle, Willim," said Jenkins of the skinny man who looked to be in his mid-fifties.

The trail soon reached the knee-deep El Tigre. Jenkins led us into the creek and we splashed upstream below ferns, philo-dendrons, and figs clinging to black canyon walls. Waterfalls plunged down side streams. The creek's waters were clear and cool, welcome in the building heat. Because the rocks on the bank were slippery, we walked directly up the sandy streambed.

Jenkins wore rubber knee boots, shorts, and a tank top: the local miner uniform. Vargas splashed upstream in his town clothes, minus the cowboy boots. He looked overdressed for wading to the waist and climbing hand-over-hand up rock walls slick with algae and mud.

Basilisks—Jesus Christ lizards—ran on the water in front of us when disturbed. Their sensational sprints reminded me of Roman at thirteen on our second trip to Corcovado, when he'd learned how to catch the miraculous creatures. First, he would chase a young basilisk across the water to the far side of a creek. Clinging to a rock, the animal would eye Roman warily until he waded too close and the reptile dove into the cool water to hide in bottom debris. Roman then reached into the mass of sticks and leaves, braving what else might lurk there, and grabbed the little lizard. Pleased with himself, he would pull it out like a trophy, inspect its dinosaur-like crest and oversized hind feet, warm it in his hands, then release it to run across the surface of the pool like some sort of windup water toy.

I tried to ignore the lizards—the memory upset me—but simply couldn't. I watched every one dash over the stream.

Our pace accelerated. I was eager both to look for clues where Roman had last been seen and to establish that Thai and I could handle ourselves here, unlike tourist gringos. We passed miners' camps as the creek threaded a series of flats and canyons. We scrambled up faint trails to rock rims above canyon slots too narrow or steep to traverse at water level. We pushed aside broad leaves adapted to deep shade and familiar as house plants in homes and offices back home.

After an hour or so Jenkins pointed out the boundary of the park. It was illegal to continue. We would face stiff penalties if the authorities caught us. "You can go back, Jenkins, but I need to go on," I said, willing to take my chances with only Jenkins's sketch map in my notebook as a guide to Zeledón. "If they find me, what can the park service say?" I asked. "I'm looking for my son. How heartless can they be? I'll take any blame." Urged on, the others agreed to continue.

Vargas and Jenkins expressed an honest parental empathy. Vargas was willing to risk his town clothes and flip phone wrapped in plastic. Both were willing to risk arrest to help, to take me where I needed to go. The young miner and the old poacher were fit and strong and knew the jungle well.

El Tigre opened up from its narrow canyons. We walked in sunshine below abrupt mountainsides cloaked in heliconia plants from creek side to ridgeline. A giant herb with wide green leaves, the heliconia displays its sturdy inflorescence of bright red and orange flowers to attract equally colorful hummingbirds. The birds in turn attract their predators: venomous eyelash vipers suspended in wait by their prehensile tails.

It's not uncommon to miss an eyelash viper with its rich camouflage. While unaggressive, the petite vipers don't hesitate to strike, and because of their arboreal habits, they often bite unsuspecting humans on the face or neck. Roman knew this and would stay clear of heliconia thickets without first clearing a path with his machete.

Piles of rocks and beds of sediment indicated active gold mining. We passed the occasional cache of hand tools—shovels and short sluice boxes—used by the Osa's miners to sift gold from streambeds. An hour into the park, and two hours from the road, Jenkins pointed out a short cliff that parted the green vegetation: "Here we must leave the Tigre on a hidden miner's trail."

We spaced out for the hundred-foot climb up a broken, slimy limestone wall with sharp hand and footholds. It led to a faint trail, choked with vines. We climbed to the rim of a deep canyon, impassable at water level, Jenkins said. Up here, a huge tree had recently fallen and buried the trail. By now it was well over eighty degrees and 99 percent humidity. We were as soaked from sweat as we would be from rain.

Jenkins unsheathed his machete and hacked a path through the claustrophobic crown of the tree fall. Each whack—and it takes many to cut through even small tropical hardwood limbs—stirs up swarms of small insects. Some bite, others sting, all leave rashes. Jenkins carved us a tunnel leading fifty yards to the twin trunks of the fallen tree. We scrambled onto its slippery surface, spooking a slender six-foot snake that slithered into the undergrowth.

Seeing me recoil, Jenkins said, "It is not dangerous." Still, I watched carefully for vipers in the foliage. The more immediate hazard was negotiating the four-foot-diameter log perched on the edge of a hundred-and-fifty-foot overhanging cliff. We balanced nervously where the log spanned a vertical gully that dropped into the precipice, then jumped off one by one, relieved to reach the narrow footpath's solid ground.

"This fallen tree is new," Jenkins said as he looked back at the log across the drop. "We have another trail that goes by my rancho. I do not come this way often."

We left the canyon rim, following a small tributary into the forest. The whole jungle was dark and dank, wet with creeks running everywhere and seeps dripping from exposed rock. The air was cool and smelled of moss, ferns, and fungi. The trail went up a shallow gully that cut into the ridgeline. In an impressive display of hand labor, miners had dug a miniature gorge ten feet deep and lined with rounded stones. It was just wide enough for us to pass, our feet splashing through ankle-deep water.

The gully opened up and we arrived at a nondescript place where the walls relaxed into slopes a person could walk up or down. Jenkins came to a stop and looked around. Filtered sunlight dappled brown leaves and black rocks. The fronds of short, waist-high palm trees swung back and forth as if in a wind, but there was no breeze. All around us katydids grated and cicadas shrieked without pause. But the two noisiest creatures of the forest could tell me nothing that I needed to know.

"This is Zeledón Creek," Jenkins announced.

CHAPTER 28

Cruz Roja

Cruz Roja and MINAE leadership, July 31, 2014.

"Here." Jenkins pointed to a pile of three rocks. "It's a little bit different now. But on that day, we were walking up this trail, just as we have been now, and sitting on this biggest rock was a guy eating his breakfast. It was in the morning before it gets too hot, maybe at eight or nine, and we had left my rancho, which is down the ridge behind us. It was me and Arley and Luiz. Coco was still back at the rancho. We were going to the mine and some tunnels we have been working. It was impossible not to pass him closely."

Indeed. There were only about four feet of level ground along the creek bottom to walk and much of that was taken up by the braids of the little creek as it spilled among the rocks.

I could feel Roman's presence here, where it was easy to see him sitting alone, enjoying the coolness of the morning, the best part of the day, pondering a line of termites on a nearby log or the curve of a tree's buttress on the hillside above while he waited for his water to boil and his food to cool. I studied the slopes and the rocks and the vegetation closely, looking for clues and wondering: *Where did you camp? Where did you descend?* And most of all, *Where did you go from here?*

Cupping my hands, I called out, "Roman! *Roooo-mann! Rooo-ooo-mann!*" But there was no answer beyond the unrelenting peeps, whistles, and buzzes of the noonday jungle.

Jenkins toured us around his mine located above the south branch of El Tigre, a creek that he called Negritos. He showed us the charred remains of the rancho that MINAE had burned to the ground, then led us down other hidden trails he and his partners used. At every turn, I studied the forest floor looking for sign of Roman: the mark of his shoes, a food wrapper, anything that might move us forward. The trails looped back to the base of the ridge at the edge of the Negritos branch of El Tigre. There we headed downstream, hiking out faster than we had hiked in, familiar now with the route.

It had been stirring to visit where Roman had last been seen. I anticipated returning soon to Zeledón. Roman was near, I could feel it. I hoped he was okay—I prayed he was alive.

AT THE CAR, I slipped money to Jenkins, who looked surprised and hesitant. He turned it down. "I have children, too. I don't need to be paid."

"Please, take it. And thank you, Jenkins," I said. "I know this was a risk for you." He turned to divide it with Vargas and Jefe sitting next to him.

"No. That's all for you, Jenkins. I'll pay them, too."

I handed Vargas an equal sum. He smiled and accepted the money in both hands, graciously.

After we dropped Jenkins off at his home and Vargas in Puerto Jiménez, Thai and I drove back to the Iguana, changed clothes, and headed to a big meeting at MINAE to debrief and discuss the search, now in its seventh day. Thirty people crowded the room: Cruz Roja, MINAE, Fuerza. Many were fresh from searching and stood in field clothes. The Cruz Roja alone had enlisted twenty-five people, nearly all volunteers. Now muddy in their faded shirts, Thai joked they should be called the "Cruz Chocolata."

A projector displayed a Google Earth image of the Osa on the wall. More than a hundred red-colored virtual pushpins showed where search teams had logged GPS locations. Three sheets torn from an oversized paper tablet were taped to the wall. Roman's missing person poster hung from one. A timeline of the last several weeks stretched across a second. Seven team names with dates next to the major trails and drainages were written on the third. Herrera, the head of the Cruz Roja, asked everyone, "Where do we go from here?"

Dondee stood up. Gesticulating to the crowd, he said that there'd been many people, and many groups, looking on all the obvious trails. He added that immediately before this search, MINAE had made a sweep of the park for illegal mining activity, adding even more information. MINAE's sweep occurred soon after Jenkins had seen Roman, when Jenkins's rancho was burned.

Dondee introduced the team leaders who'd combed the trails

and logical routes across the hundred square miles of Corcovado. They rarely found signs of other teams who had passed only days before; rain and litterfall had erased their tracks. Cruz Roja and MINAE, escorted by Fuerza, had looked in every drainage mentioned in Roman's emails. They had even struggled across the disorienting plateau known as Las Quebraditas, visiting its high points Mueller and Rincon, marked by geodetic benchmarks, which are metal disks embedded in concrete pads.

The four-day search through Las Quebraditas' cloud forest was led by a thin young man in a floppy sun hat who would be my first pick for a search team—if Dondee would let me assemble one. His team had followed a tourist trail on a two-hour approach to Las Quebraditas. The trail then thinned to an unmaintained trace following a poacher's trail. For six miles, they struggled to find and hold the faint track, crisscrossed with fallen trees that deflected them into thickets of bamboo riven by shallow, slick ravines. Each step off-trail reminded the team members how easily they could lose their way.

Beneath a canopy too thick for GPS signals, and in vegetation too tangled to follow compass bearings, the searchers found themselves wandering in circles. The only tracks they saw were their own. The challenge was that the most likely place to become lost was also the most difficult to search.

Not a single team unearthed any sign of Roman. Superintendent Arce pointed out that some places were just too steep and illogical to look.

After the seven teams briefed the assembled group, Dondee turned to me. In a challenging voice, he said that he knew that we had gone illegally into the park that day. "Who were the three others in the car with you?" he asked accusingly. "And where did you go?"

To protect those who'd helped at great risk to themselves, I withheld Vargas's name and claimed that Jenkins walked us to the park boundary but no farther. Thai and I had pushed on alone, I lied, to visit the point last seen. While all of Costa Rica and the U.S. wanted to help, a poacher and an illegal miner had been most effective. I wasn't about to give them up.

Dondee saw through my lie and scowled. Adding injury to insult, he finished by saying the search would end soon. MINAE would continue to look for sign as part of their ongoing park patrols, but, in his opinion, Cody had never entered the park in the first place: he had been with Pata Lora on the Piedras Blancas trail.

CHAPTER 29

Whiteout

Above El Doctor, August 1, 2014.

Dondee's opinion mattered as much to me as mine did to him.

Cruz Roja would officially suspend the search three days later. While society's expectations might then be met, mine would not until my son was found. A parent's loss goes so deep that they'll do anything, stand up to anybody, to save their child. Dondee knew this and warned us that there'd be people waiting to arrest us in case we searched the park illegally, especially if we paid people who were not licensed guides, further violating the park's regulations.

Based on Jenkins's and his two partners' descriptions and Roman's own emails, I was certain Roman had entered the park and been seen on Zeledón Creek—maybe as recently as sixteen days before. He could still be alive somewhere upstream. And instead of thirty people with computerized debriefings and thousand-dollar-an-hour helicopters, why not a focused search by Vargas, his son, Thai, and me, starting near Roman's place last seen on Zeledón? We could head upward from there, into Las Quebraditas, the most likely place to be lost and injured. We made plans to meet at six the next morning.

ANOTHER SEARCH FROM thirty years before emboldened me. It was a time before specialized teams of roped rescuers existed in Fairbanks and it was up to us in the climbing community to bring home our own. One night we got a call from the Alaska State Troopers who said there had been a climbing accident in the Hayes Range. Our community was small. We all knew it was Carl Tobin or Matt Van Enkevort climbing Ninety-four Forty-eight, a mountain Carl and I had failed on three years before. As more information trickled in, we discovered that Carl had been seriously hurt during a long fall in an avalanche. Leaving Carl in a small bivouac tent, Matt had skied and hiked twenty miles until he found a moose hunter's camp with a radio and used it to call for a rescue. It wasn't clear yet what the extent of Carl's injuries were, only that he had broken both legs, including his femur. Carl was my regular climbing partner and I feared he could die.

The next morning, four of us left before dawn in an army helicopter that dropped us at the toe of the Gillam Glacier. The Range was swallowed in storm, but we headed up-glacier anyway. We pulled a big sled to bring Carl back. Whiteout had reduced visibility to a few yards and forced us to follow the wind's direction calibrated by compass. Never sure where we were, we knew only the direction we moved.

Tied together, we skied into a wind so fierce it knocked each of us down at some point. Miraculously the storm slackened and a hole in the blizzard opened that reached across the glacier. Through this window I saw we were near Ninety-four Forty-eight, and with the improved visibility, I spotted Carl's tent above the glacier on a moraine, a low ridgeline of rocks left by the glacier's movement.

The break in the weather had come at just the right moment. Realigning my compass, we skied into the teeth of the wind as the hole in the storm closed again. First on the rope, I fought off my fear for the worst as we skied up the moraine. But as we pulled to its top, we saw that it wasn't Carl's tent, but a boulder. Dismayed, I feared we might not find him at all. The glacier was big and the whiteout hid everything more than fifty yards distant.

I led us to the boulder anyway and looked downhill past it, seeing behind the moraine for the first time. And *there* was the tent! We hurried down the moraine. My mind raced. *Will we find Carl alive? And what if we don't? What then?* The tent flapped wildly in the wind, but had a strange, counter beat to it, too. As I closed in on the half-buried shelter, I heard loud cursing.

"Carl!" I yelled. "*Carl!*"

"Yeah!" I heard Carl's voice from inside the tent, "Yeah. Hey, who's with you?" he asked.

We were all relieved he was alive. It seemed incredible, given the conditions, that we had found him at all. We secured Carl in the sled and worked him down the glacier through the storm and whiteout to the helicopter pickup. I thought about how we had found him. We were a small group of his friends who had the skills and knowledge to know where to look and how to get there, coupled with resources, like the army's helicopter, for support.

But the real lesson had been this: *Follow intuition. It often leads in the right direction, if not directly to the destination.* If we could find Carl on the Gillam Glacier in a whiteout, then we could find Roman in the jungle.

CRUZ ROJA'S ANNOUNCEMENT that they would soon call off the search had only galvanized our resolve to search on our own. Thai and I shopped for three days of lunch food to supplement the freeze-dried dinners we had brought from Alaska. We packed light: a bug tent with a tarp-like rain fly, stove and a cookpot, sleeping pads, sleeping sheets, dry clothes, and head lamps, all wrapped in dry bags inside our packs. We would be ready to leave at dawn.

That night at the Iguana, as I tried to sleep, my phone rang with an unknown number. The caller was cagey, but offered to help. "How?" I asked.

"Tell me what's happening."

I gave him the story, ending with how the Costa Rican Red Cross had kept us out of the park.

"The Red Cross sucks," he said. "But I can help. I hear through the grapevine that a black snake has your son."

"A black snake?"

"Yeah, a bad motherfucker. That's what we do. We deal with black snakes and do extractions. I have an asset in Costa right now. It usually costs thirty but we'll do it for fifteen."

Unsure what I was hearing, I stuck to the facts. "Well, I'm going in tomorrow. Somebody saw my son, talked to him a couple weeks ago. I went in there today, where he was last seen, and we're going back."

"Who are you going with?"

"A local guy. He knows the place well."

"Do you have anybody to watch your back?"

"Watch my back?"

"Yeah. Can you trust this guy? What do you have for a weapon?" That got my attention.

"Um, no. No weapon. But I have another friend from the States with me."

"Oh, okay," said the cagey voice. "Look, I'm going to text you my number and if you need help with the black snake, get a hold of me."

Then he hung up.

What the hell was that?

CHAPTER 30

Las Quebraditas

Vargas in Las Quebraditas, August 1, 2014.

Early the next morning and supplied with food and camping gear, Thai, Vargas, his son, and I retraced Jenkins's route to Zeledón. Hungry for anything that might help me find Roman, I lingered and searched for clues at the nondescript boulder where he had eaten breakfast weeks before. If he'd come this far, he would have likely kept going, toward the disorienting jungle of Las Quebraditas, where we would head to next.

About 150 yards beyond, we took the right fork at a split in the trail, following a faint trail to the ridge crest. The day before,

Jenkins had taken us left to mining tunnels punched into a canyon wall above El Tigre's Negritos branch. Roman had indicated to Jenkins he would continue onward. Using the rule he'd developed in Mexico and Guatemala, he likely chose the better-used left fork, well worn from the four miners' daily commute to their dig.

Vargas led us on the right trail that followed a narrow ridgeline falling steeply to either side. I studied the trail for footprints of a Salomon shoe, the kind I knew Roman wore. I called down into the canyons on either side: "*Ro-man! Ro-mannn!*" Jefe did likewise. Thai blew a loud, fist-sized rescue whistle. Only echoes responded.

SHARP, HAPHAZARD MEMORIES of Roman crowded themselves into the search. The time he came home from school announcing "Dad, let's play chess" came to mind. He was in his mid-teens then, during our golden age together. From his room, he brought out an ornate Balinese box, filled with hand-carved chess pieces, that unfolded as a chessboard. He assembled the board quickly, then held out his hands, a pawn in each. I chose black and he went first. His moves were swift, decisive. He beat me handily. Bewildered, I said, "Wow. Good job. Let's play again."

"Okay," he agreed, beating me a second time.

"You're pretty lucky," I said. But when he beat me a third time—grinning from across the table—he said, "That's not luck." I had to agree. He was good.

It was one of those moments that marked his growth, like when he first stood up as a toddler in the little house in Fairbanks, or told me from the Mexico City airport that he was going to catch the last bus. I prayed this trip in Corcovado would also be a step forward for him, and not an end.

THIRTY MINUTES BEYOND Zeledón, a fetid odor hung in the humid air. Fearing the worst, I left the trail and found the rotting carcass of a tamandua, the small black-and-cream-colored anteater that lives throughout Central America. Thai and I had seen one alive the day we drove to Carate. Roman had an interest in the group of odd, New World mammals—the sloths, armadillos, and anteaters—classified together as the Edentates. I had hoped then that the roadside tamandua was a good omen.

We followed a subtle trail used by poachers, illegal gold miners, and the park rangers who hunted them both. Just a narrow wisp of a path, it was something most hikers would lose quickly or dismiss as an animal trail. Only the occasional machete nicks on heliconia plants, palms, and ferns showed it was actively cleared. It wasn't blazed at all.

Small two-by-two-foot clearings on the ridgeline marked the rare spot where locals got a bar or two of cell reception. Vargas stopped at one, unwrapped his flip phone from a small plastic bag and called his daughter. It would rain soon and he wanted to let her know we were okay. Thai smiled at me and motioned toward Vargas. "It's like a little jungle phone booth," he joked.

We would camp our first night in the heart of Las Quebraditas on the Osa's summit plateau. The place was a veritable maze of bamboo-choked gullies in a remote mountain wilderness. No wonder the Cruz Roja team leader in the floppy hat had been confounded here. Without Vargas, we, too, would have been circling back on our tracks.

The trail thinned. Vargas left it and we wandered through bamboo thickets and slick gullies, looking for a place to camp with running water. The rain caught us before we found a spot flat enough for tents. Soaked, Thai and I set up our fly, then erected the bug net tent beneath it, keeping dry even in the pouring rain. It felt comfortable to get out of wet clothes and into dry ones and tuck in under our sheets.

Meanwhile, Vargas and Jefe unsheathed their machetes and made camp. First, they hacked a beam and tied it to two trees to hang their Visqueen tarp to get out of the rain. Next, they cut and assembled bedposts, a frame, and slats to sleep in a hand-made bamboo bed, three feet off the ground and out of the way of snakes, ants, and spiders. They used fern fronds as a thin mattress. Finally, a smudge lit under their tarp kept down the bugs while the two slept out in the open air.

In the morning, we set off into a flat jungle dense with vegetation. The sky was overcast by ten. Without the sun as a compass, I soon found myself lost, failing Vargas's test when he asked us the direction from which we'd come. I pointed one way. Thai another. Chuckling, Vargas indicated a third. Wandering between the featureless summits of Mounts Mueller and Rincon, we found ourselves holding tight to Vargas's lead. At times, even he seemed uncertain, slicing his way up and down trails marked only by the hoofprints of peccaries.

Dropping into the namesake little gullies of Las Quebraditas, we plunged hard on our heels to anchor our feet and tried not to grab the stems of spiny palms covered in inch-long needles. I leaned on a trekking pole for balance but wished for a handrail to keep upright. Every so often and with a single swing, Vargas would hack a bamboo stalk as big around as my arm, then offer us a drink of sweet, cool water from inside the hollow stem.

Around midday, Vargas led us down out of the claustrophobic bamboo forest onto a broad ridge of open rainforest stacked with buttressed trees and an understory of philodendrons. He cut two-foot heliconia leaves as seat covers against the ants and the fungi on logs where we sat for lunch.

Thai translated Vargas's thick country accent with difficulty, often asking Jefe what his father had said. Thai pointed down the broad ridge where a tapir or maybe a hunter had passed. "Vargas says this is the way down to the Rio Claro."

I wasn't ready to go that way yet. Given the maze of Las Quebraditas and the subtleties of its trails, it seemed unlikely Roman would have made it through this keyhole leading off the summit plateau and down to the Pacific. I studied my topo map. At the top of the Osa, Mount Mueller forms the center of a five-pointed star with each vertex pointing to a different drainage: three to the Pacific and two to Golfo Dulce.

I asked Vargas to take us directly down to a tributary of the Piedras Blancas arm of the Rio Tigre and back to Dos Brazos where we had started, closing a big loop. The steeper terrain and thinner vegetation would naturally draw a hiker in that direction. If Roman had followed the line of least resistance from Zeledón, then east-west-trending canyons would have funneled him toward Las Quebraditas.

Relying on intuition and compass bearing alone, and given his experiences in El Petén, Roman would have balked at pushing through Las Quebraditas' confusing landscape. It seemed more prudent to search closer to Zeledón Creek, where Roman was last seen, than here on the far side of a maze.

Dropping off Mount Mueller's slopes there were no human, no tapir, no peccary trails. Just raw jungle travel. Even Vargas hesitated. He was tense off-trail, haunted perhaps by the memory of the bushmaster that bit and killed his brother on the spot. The bushmaster is not only the longest poisonous snake in the Americas, but also its most aggressive. Once agitated, it rarely backs down.

Vargas plunge-stepped down the steep slope of mud. It was hard to keep up, even while he sliced the herbaceous growth with his machete as he bushwhacked off-trail. The blade let out a reassuring *tzing, tzing, tzing* that left a path clear of snakes and a trail of fresh green leaves to follow should we need to retrace our steps, like Roman had in El Petén.

Midway down, a series of waterfall drops forced us to lower our packs to each other. The exposure here emphasized how readily Roman could have been injured had he slipped into a steep gorge or canyon, like those in the Negritos below Zeledón. I vowed to return to Zeledón and search Negritos's canyon with ropes and climbing equipment.

The travel was difficult, not physically but emotionally, especially when calling his name. My grief painted the jungle black, but the heart of the Osa's wilderness still left me awed. Every neon-colored dart frog, every emerald green bird, every fascinating jewel of the jungle that we passed left me with a pang of regret and sadness, remembering how our family had thrilled together at rainforest wonders. Those vivid memories grimly reminded me of why I was here. They left my eyes watery, my heart heavy.

But I couldn't shut out forever the joy in seeing a kingfisher's blue flash or a spider monkey's graceful swing. To ignore those pleasures devalued our lifetimes shared in places like this where we marveled at nature's creations. Sometime on the third day, I could again see rainforest colors and delight in the flight of a basilisk across a stream or the primeval look of a motmot in bamboo.

After we made our way back to the network of miner trails, Jefe killed a small fer-de-lance at an abandoned miner's camp. Young poisonous snakes are the most dangerous. In their youthful inexperience, they have not yet learned to regulate their venom's delivery, often over-envenomating in self-defense. A sixteen-inch juvenile can readily kill a man.

Soon after, Thai stepped over a log where, unknown to him, an olive-green eyelash viper was coiled for a strike only inches from his femoral artery. He could just as easily have put his hand on the snake, or swung his leg over and sat on it.

Thai was five strides ahead of me when I called out to him, "Hey, *Thai*! You just about got bit by a viper coiled on this log!" I held out the little green serpent, its prehensile tail wrapped tight around my trekking pole.

Thai just flashed that world-wise smile, shook his head a few times in disbelief, then turned and hurried through the heat back to town where the Cruz Roja was closing down the official search for Roman.

CHAPTER 31

Negritos

Steve Fassbinder rappeling a Negritos waterfall, August 11, 2014.

The official search was over. Weary Cruz Roja volunteers headed home to their jobs. In a meeting, Dondee reminded Thai and me that my son had planned to enter the park illegally, that searches for illegals were difficult to approve in the first place, and that an exception had been made for him. Resources needed for other searches had been consumed here. Cruz Roja and MINAE would not resume their search without hard evidence.

Dondee also discouraged any offer of a reward, bringing to

mind the painful story of David Gimelfarb. In 2013, four years after he had disappeared, his parents received several phone calls. The caller claimed a drug cartel held their son hostage. For twice the reward offer, the caller would reveal their son's location. FBI investigators know that Latin American criminals take advantage of families of missing persons and dismissed the call as a scam. There was little chance anyone would hold a hostage so long before asking for ransom.

The same day Dondee said was Cruz Roja's last, Thai went home to his wife and infant daughter, leaving me depressed and alone. Sitting at the Iguana Lodge with my hands tied, knowing every day counted more than the last, I sobbed briefly, as I did every day during private moments. I quickly choked back to regain control. Guilt followed.

What kind of father have I really been?

Parents aren't supposed to pass out pills, smoke dope, or drink booze with their kids, and we never did. Instead, we bought them airplane tickets to exotic lands. Travel itself can be an addiction. Adventure is. Here I was, searching for Roman missing on a trip that traced directly back to me. I had not simply *introduced* him to international travel and the risks of wilderness adventure. I had *included* him, again and again, to the point that a large part of our relationship—his very name—was built on experiences like his illegal bushwhack into Corcovado.

I couldn't shake the feeling that everything I had done with him in the wild had all been a mistake, that in the end, *I had been* that irresponsible father the cowboys saw on Umnak. I might not have hurt the six-year-old boy then, but the suffering of a twenty-seven-year-old man lost and broken in the jungle now felt like my fault. Yet every time those thoughts circled round me, Tennyson's words came too:

I hold it true, whate'er befall;
I feel it when I sorrow most;
'Tis better to have loved and lost
Than never to have loved at all.

The love that I had for Roman—and for Peggy and Jazz, for that matter—was stronger and deeper for the time we had spent together in wild places. I would not give that up, even as I felt more helpless than ever. And while moments like this would plague me—still plague me—I would hold it as a truth that the bonds we form in nature with others are the truest bonds between us. While Roman may have been lost and dying because of our time in Australia, Borneo, or wild Alaska, that time we had together compelled me to come and do whatever was necessary to find him now.

SOON, THAI'S FRIEND Ole Carillo from Anchorage; my friend Steve Fassbinder; and his Spanish-speaking coworker, a young woman named Armida Huerta, both from Colorado, would arrive. Ole lacked Thai's wilderness skills, but he was even more easygoing and had nearly as much travel experience. He also spoke fluent Spanish. I knew Steve well from a two-hundred-mile beach bike and packraft trip along Alaska's southern coast, but I'd only soon be meeting Armida. Neither had tropical experience.

Meanwhile the Pata Lora story had seeped deeper into the Osa, into every pulperia and hovel. A rumor had spread that we were offering a reward. The situation was spiraling out of my control. But a lifetime of risk had taught me that a calm mind works better than an excited one. On this—the most important journey of my life—I controlled what I could: my emotions.

Dizzy with a pounding headache, I woke sick to my stomach

and hurried to the toilet. Chewing Pepto Bismol pills only added nausea to my diarrhea. Morning meetings with officials didn't make me feel any better; they thwarted my plans to ask the Cruz Roja for personnel and a long rope to take into the Negritos. Park Superintendent Eliecer Arce, a father himself who was sensitive to my plight, remained adamant that the area was illegal for anyone but park officials.

I decided to keep my canyon rappelling plan to myself. Every official made it clear they were upset with me already, both for my known forays into Corcovado and for the ones they suspected I'd made or soon make.

Back in Alaska, Peggy fielded phone calls and sifted through offers from Facebook friends. Most were of the we have contacts in Costa Rica variety:

I just heard a little about your son. Interestingly enough, my next door neighbor here has a nephew who owns a place called Good Times, a surfer retreat in Costa Rica. He has been there a while and speaks Spanish. If you give me information, I can pass it along and maybe he can do some nosing around for you.

So many people *wanted* to help. But people asking questions around the edges of Corcovado, Puerto Jiménez, and Carate would simply turn up the Pata Lora story. We needed immediate assistance from people with tropical search and technical rescue skills. Mead Treadwell and his friend Josh Lewis—both active Alaskans in the venerable Explorers Club—were eager to effect this kind of assistance. Mead even took precious time out of his run for U.S. senator to do what he could.

Mead wrote a letter of introduction to Costa Rican officials that described me as "well-known for exploration and search and rescue work under very difficult conditions in many climates including tropical rainforests." He informed the embassy that I was "more than a distraught parent," but "an asset [that he] would want on any search in these conditions." In the end, MINAE permitted me into the park because of Mead's efforts and Josh Lewis's connections.

The son of a successful Colorado oil man, Josh had an old family friend in Costa Rica named Juan Edgar Picado, a lawyer in San José. Juan Edgar's father had been very influential in Costa Rican politics, as was Juan Edgar himself. Juan Edgar was a personal friend of Costa Rican president Luis Guillermo Solís and of Public Security Minister Celso Gamboa Sánchez, the Costa Rican equivalent of the U.S. secretary of defense.

Minister Gamboa signed a letter giving me and my friends special dispensation to enter the park. As part of the deal, we had to fax copies of our passports and signed, notarized statements to San José waiving Costa Rica of any responsibility should we be hurt or killed. We were also required to travel with a MINAE permit and rangers.

Impatient to get back to the jungle as soon as possible, I wrote the GPS coordinates of Zeledón on the documents, faxed them to MINAE, and suggested they send our permit in with the ranger when it was ready. Ole, Steve, and I left for Dos Brazos without the permit or rangers. Heeding Dondee's threat of arrest if I were caught, I shaved off my beard and ducked low in the back seat when police, MINAE, or Cruz Roja vehicles passed by. It was discouraging to think that officials might now put more effort into stopping me than looking in the park for Roman.

A multisport athlete and adventurer, Steve carried with him a pair of two-hundred-foot ropes and climbing gear to rappel into the Negritos canyon. Without ropes, the canyon is inaccessible, blocked by waterfalls at top and bottom. We hiked into Zeledón, set up camp on one of the few flat spots above the creek, then scrambled down to the Negritos branch of El Tigre where it drops off the first of a half dozen waterfalls. Jenkins said Roman had climbed out at the lowest waterfall. I thought that perhaps he could have somehow fallen into it afterward, in a Hollywood version of a slippery slope in the wilderness.

It was early afternoon and clouds obscured the sun. Rain was coming. We set up the first rappel to slide down the rope into a bowl carved from pebbly walls. "Ole, have you done much climbing?" Steve asked casually.

"Nah, not much," Ole replied, "but I have rappelled before."

"How about climb a fixed rope with ascenders?"

Smiling, he shook his head. "No, I haven't done that, I'm afraid."

Steve and I stretched out the long rope and dropped off one waterfall and then another. Ole followed. Steve scouted deeper into the slot canyon downstream. By the time he had returned it was raining hard. Steve yelled over the din, "It'll go with more rope! But I think we should get out, *now*!"

Steve had seen enough flash floods to know when to go. By the time we all had climbed out of the gorge, the creek had risen to an uncrossable depth. We got out just in time. Clawing our way up the greasy canyon walls and worried about these two friends in a dangerous place, I understood now the park officials' and Cruz Roja's concerns about me.

The next morning, Ole stayed at camp. Steve and I rappelled Negritos's waterfalls, cutting our rope at the bottom of each drop so we could climb back up with ascenders. Between waterfalls, we scrambled and swam the Negritos as it slithered through slot canyons coated in green algae and moss.

Each waterfall was choked with logs and wood. Midway through, we found a broken machete, its rusty blade thrust into a log spanning the creek. It looked too old for Roman to have left it only weeks before. Other than the machete, we found no sign that anyone had ever been there. Below the last waterfall, the walls barely kicked back enough to exit the canyon, just as Roman had described to Jenkins when they had met.

By the time we left on our third day, I was convinced Negritos's waterfall-filled side gullies and its upper tributary El Doctor needed thorough searching, too. Zeledón itself was perched between the Negritos and another branch of El Tigre that we also had not explored. It was just damned hard to get there, bushwhacking through all the red tape.

CHAPTER 32

Piedras Blancas

Roy Arias's house, Piedras Blancas, August 10, 2014.

While we searched Corcovado's canyons, Armida Huerta interviewed people in town. Most told her the Pata Lora story. But she did hear a new one. Sean Hogan, an American living on the Osa, described a gringo he'd met on a weekday morning in Puerto Jiménez around July 7 or 8. The gringo "looked similar to the photo on the poster, but was more tanned, a bit thinner and older too, in travel-worn clothes, like he'd been out for a while." The young man Sean met was quiet and didn't volunteer much. Instead he'd asked questions of Sean. That sounded like Roman to me.

Lauren picked us up in Dos Brazos near noon. Back at the Iguana we sat down with Josh Lewis and his wife, Vic. The families of Josh and Juan Edgar Picado were united by the Fellowship, a Christian political organization based in the United States but international in scope. I'd soon be impressed by the reach and effectiveness of this group and grateful for its efforts to get search-and-rescue personnel from the U.S. military involved. Josh and his wife had flown down from Alaska to help. With his big white beard and aloha shirt, he looked like Santa on vacation.

Josh had employed a Tico driver from San José. Over lunch the driver earnestly recounted the Pata Lora story that he had heard at a bar the night before. I rolled my eyes and tried to explain *that* was not my son. The driver's look alone was enough to say: "This father sees his son through rose-colored glasses."

On the phone Peggy told me, "You should question the people who say this Pata Lora guy was with Roman. We know he lied about the ATM. We should at least find out why."

I had no doubt at all that Pata Lora had made the Dos Brazos–Carate crossing with a gringo. Multiple people had seen them together. What I needed to know was if Pata Lora's Cody was our Roman. It was time to visit Piedras Blancas, midpoint along the "Pata Lora trail" from Dos Brazos to Carate.

At the heart of Piedras Blancas is its only permanent structure, a two-story house occupied by Roy Arias—a responsible miner, according to Lauren. Pata Lora and Cody had visited with Roy on their way to Carate, even camping near the house. The Iguana Lodge's gardener, nicknamed Chico, agreed to lead us there. Chico's own father had been killed by a *terciopelo*, the local name for the velvety-skinned fer-de-lance, when Chico was a child.

After four hours of hard walking, we arrived at Roy Arias's place. White ponies grazed in a grassy pasture outside the open-plan house. Inside, several hammocks hung between posts; colorful laundry dried from a line on the first floor. We questioned several Piedras Blancas miners whose stories roughly matched those of witnesses from Dos Brazos. My notebook recorded Luis describing Cody: older than thirty; yellowish-brown hair; no glasses; no beard; more hair than me that was combed back; wearing Crocs; and smoking pot that he pulled from a big satchel.

Next, we tracked down Roy Arias. Wearing the miner's uniform of knee boots, board shorts, and a cutoff T-shirt, Roy looked to be in his forties. He was digging at a placer gold deposit with his partner, Chelo. The two worked the stream by muscle alone, prying pay dirt with an old shovel and sifting gold in a three-foot sluice box. I showed them recent photos of Roman. They laughed and nodded favorably at the photo where Roman posed next to a bikini-clad friend pulled close to his side.

Ole translated. "Roy and Chelo say the guy with Pata Lora was not the guy in the photos. He didn't look like him at all." The miners said the gringo with Pata Lora spoke little Spanish; didn't cook; slept in a tan tent; and wore Crocs. The part about the Crocs didn't resonate with me.

On the APU trip to Corcovado when we'd walked across the park, eleven-year-old Roman had witnessed a sandal-wearing student impale himself on a six-inch palm spine. Carl and I had sent the student out on horseback to go to the Golfito hospital. To clear his subsequent infection, the student required an antibiotic drip that he carried in a fanny pack for six months. While there may have been a Cody walking in Crocs with Pata Lora through Piedras Blancas and on to Carate, it wasn't my Cody Roman.

BY THIS TIME, a month had passed since Roman should have returned from his Corcovado crossing. Knowing the odds and desperate to try anything, I resorted to checking the geographic coordinates supplied by a psychic. The position was off-trail near Sirena, Corcovado's chief tourist center and a full day's walk from Carate. Accessible only by foot, boat, or air, the flight to Sirena from Puerto Jiménez is expensive but short; we flew because it was quick.

Ole, Steve, Armida, and I needed a licensed guide to join us in the park. The pilot recommended young Nathan. When Nathan learned of our plan his eyes went wide. "We can't leave the trail," he declared. "I'll lose my job and probably my guiding license." Finally, at risk to his livelihood, Nathan agreed to lead us along the trail. Family is important in Costa Rica, and everyone we'd met was willing to do what they could. Everyone wanted me to find my son.

The flight to Sirena traced an achingly beautiful coast where the jungle tumbled down to the sea. I looked over a geography I knew well but for all the wrong reasons. Fifteen minutes after leaving Puerto Jiménez, the Cessna bumped in landing on the grassy strip. Lunchtime tourists packed the boardwalks and platforms in clusters, each with a young man like Nathan who toted a scope on a tripod and pointed out monkeys, toucans, and sloths in the trees.

Fifteen years before, when eco-tourists were few and no guides were required, Roman and I had walked with the APU class from Los Patos to Sirena, where we stayed for several days. In the forest, Roman staged battles between ants and termites, mortal enemies in the war for the jungle.

"Who wins?" I asked him.

"The termites put up a good fight with their nozzle-headed soldiers shooting goo," he said, "but the ants always win. They have more soldiers."

Roman even nosed around a storage building and caught a

nectar-eating bat. There are few animals as exhilarating to hold as those that can fly. While Roman held it, the bat's tongue darted out investigating and probing his gloved hand. He marveled at its long, skinny pink tongue, used to slurp nectar from tubular white flowers that open only at night.

There was a tame toucan there, too. Roman had touched its yellow-and-chestnut-colored bill. "What's it feel like?" I asked.

"It looks heavy and solid, but it's not. It's hollow and light."

So vivid were the memories at Sirena that I walked behind my friends, wiping my eyes in private.

Near the psychic-supplied GPS coordinates, Nathan looked up and down the trail. When the coast was clear, he whispered to us to leave the ten-foot-wide tourist trail and head into the lowland forest. Less than fifty yards off-trail, we encountered a dozen peccaries. The size of pit bulls, the wild pigs were curious and nearly touched us, their twitching snouts sniffing our knees.

I looked hard for Roman's Kelty tent with the navy blue fly he'd brought from Alaska. *What will I say to him? What are best- and worst-case scenarios? Why has he ended up here, of all places?*

The peccaries followed us for twenty minutes through ankle-deep water beneath head-high palms and an overstory of tall buttressed trees. Peccaries, like all pigs, are omnivorous scavengers. I couldn't help but imagine the worst. We wandered deeper into the muddy forest, but other than palm foliage pinned to the mud by fallen branches, nothing was disturbed. No sign, no footprints, no tent, no stink, nothing but another dead end.

CHAPTER 33

Homefront

Kitchen, Anchorage, August 2014.

The Kübler-Ross model postulates five stages of grief: denial, anger, bargaining, depression, and acceptance. These feelings swirled interchangeably in me for weeks. Every Tico and Tica suggested strength—*fuerte*—to cope with my grief. I did my best to do normal things: write, take pictures, tell stories, laugh. Still, almost anything could trigger a memory, sometimes so strongly that a wave of grief swelled, crested, and crashed over me. I'd weep for an instant, then get back to work.

After Sirena, I called Peggy from the Iguana to tell her I was leaving the Osa to go to San José, where Josh and Mead suggested a media campaign to drum up military support. Just hearing her voice, sweet and present, revived me from the dead ends in Negritos, Piedras Blancas, and Sirena. She never sounded down or depressed: only upbeat, empathetic, supportive, and loving. Peggy gave me the most fuerte of all.

Back home in Alaska, she faced struggles so much harder than mine. At least I could do something firsthand rather than rely on the actions of others. Unlike me, Peggy answered phone calls and emails, coordinated help and support. She communicated with everyone from reporters seeking a story to strangers offering to help. It was a full-time job.

Expenses piled up, too, with food and lodging; car rental; international cell-phone calls; logistics for friends coming down to help; hiring guides. Peggy managed the contributions that family, friends, former students, and even generous strangers gave us to help pay for all the costs.

The 2014 Wilderness Classic began while I was away. I'd planned to race with a friend in a two-person packraft and boldly descend the Tana, a major glacial river near the end of the route. Tragically, one of the 2014 racers, a well-liked, good-natured, and experienced veteran of the race, died on the Tana when his raft flipped in an icy Class IV rapid swollen with glacier melt. In spite of many close calls over the race's thirty-year history, his was the only death ever in the event. The racers had all donated their entry fees for Roman's search before the race start. Peggy said the dead racer's check sitting on our kitchen table was a poignant reminder of the sometimes irrevocable cost of adventure.

Like a mother bear, Peggy squarely faced any threat to her offspring. With her phone always close and posted at her computer day and night, she dispensed news from me and answered the same questions from others over and over. We shared emotions that peaked when we were convinced that our son was alive and well (just ignoring and avoiding us), and plummeted when we imagined him lost, injured, suffering, or worse.

Peggy took care of details that could only be handled from home. Because we'd sent Costa Rican authorities an outdated picture of Roman from 2012, she searched for more recent ones: with his friend Denali in Hawaii; at home with his sister in Anchorage; on a boat fishing in Alaska's Prince William Sound with Katelyn; in Guatemala with traveling companions. His straight white teeth showed behind a broad smile in each. Peggy forwarded these to Costa Rica for distribution to Fuerza, Cruz Roja, and MINAE.

In a little-known international agreement, the National Guard of every state in the U.S. is paired with an American ally to help in humanitarian crises. Lieutenant Governor Treadwell suggested that the Costa Ricans pull the New Mexico National Guard into the search. Peggy urged Alaskan politicians to follow up on this suggestion.

She also pleaded with bank representatives to share Roman's last financial activities. "If there's one thing to be learned from this," she told her friends, "it's to be sure that someone else is listed jointly on your children's bank accounts. Otherwise you'll never be able to track their financial movements if you need to find out where they were last."

The office of Alaska's then senator—Democrat Mark Begich—called Peggy to say the senator sometimes involved himself in missing persons cases. However, while neither he nor his office ever said so, it seemed to us unlikely that a sitting Democratic senator, up for reelection, would pitch in with Treadwell, a Republican campaigning for Begich's seat in the upcoming November election.

All officials, from the embassy, to the FBI, to the senators' offices, asked the same missing persons questions: did Roman have a Facebook page, a cell phone, a GPS; did he use drugs; how experienced was he; when did we last hear from him; etc., etc. Peggy answered the American authorities with the same responses Cruz Roja got from me my first day in Costa Rica.

Like Roman, Peggy had avoided Facebook. But now she found it an efficient tool for connecting with and updating people. In an outpouring of support, Facebook friends reported that family, friends, and former guides were willing to help. But what Facebook friends couldn't know was that even an army of well-meaning acquaintances and friends of friends would have no better luck getting permission to enter Corcovado than we had.

Even if they did get into the park, how many Facebook friends had the jungle savvy to follow thin, unmarked paths used by poachers and illegal gold miners, trails meant to be hidden? How many could dodge green vipers at eye level and fer-de-lances underfoot while watching their step through slippery mud without grabbing spiny palms as handholds? Jungle travel needs four eyes and a sixth sense for hazards. How many had those and the time to come down, even if our GoFundMe campaign could foot the bill?

At first, Peggy kept a list of these names and their contact details. Some offered places to stay, locals to translate. But as our private search efforts were both illegal and risky, she stopped keeping the list and instead politely thanked those who offered. By two weeks in, the sleepless nights, incessant communication, and stress over her missing son had just plain worn Peggy out. Thinking about what her son was going through—out of food and three weeks overdue—made it harder and harder to keep it together.

She confided her fears to her friends, but never to me. She was confident Roman was alive, and I needed her faith. Early on in the ordeal she had broken down and sobbed long and hard, flushing grief from her system to better focus on the tasks at hand. Friends invited her to get away from the phone and the computer, to berry-pick, or just walk and talk. They gave their prayers, their love, their money. They shared stories about their kids as distraction.

Peggy's brother-in-law Steve, together with sister Maureen, Carl Tobin, and other friends and neighbors, completed a house siding project left unfinished when I dropped everything to head south. Steve and Maureen helped Peggy strip and sand our living room floor. Peggy sent me photos of their work. It all looked great and provided her a constructive diversion from worry. "Keeping busy is good," she wrote. "It helps keep the breakdowns at bay."

AS EARLY AS July 29, a slew of high-ranking politicians—Treadwell; Alaska senators Lisa Murkowski and Mark Begich and Congressman Don Young; Florida senator Bill Nelson; and a handful of generals including General John F. Kelly—all expected that National Guard personnel would soon be deployed, like a cavalry to the rescue.

But by the middle of August, Peggy was fed up with the National Guard runaround. All the back and forth, sucking up to the press, begging people to write their congressman, enlarging the circle of contacts to see who might finally get the message to President Obama for permission to send a small group of trained rescue personnel—like the Air Force's PJ rescue squad—had stretched from days into weeks. All Peggy and I could see was the ticking clock.

"I'm starting to ignore this shit and told Roman to just get back into the jungle and fuck em all. If alive, my son is presently dying, or dead from all this bullshit," she wrote, venting to a friend, sick of the empty promises and broken dreams. Frustrated at the pace of action, she fired off a scathing email to Begich's office:

> We still have not received capable people with jungle/rope skills—something we have been striving for three weeks. We feel so close, then get pushed back ten steps. I'm not sure what Mark has actually done himself, but it sounds like he could do more. ONE TELEPHONE CALL OF HIS TIME. PLEASE.
>
> The Republican Party is shining bright right now. Really Bright. Extra Bright. I'd like to see Mark kick it up a notch.

That worked. Begich called me the next day. But every politician's promise from D.C. to San José delivered little more than the false hope I'd found in the psychic's GPS coordinates near Sirena.

CHAPTER 34

The Fellowship

Juan Edgar Picado, San José, August 2014.

Back in Costa Rica, I joined Josh and his wife, Vic, in San José, where I felt like a sad pony trotted out for sympathy. We hoped that Juan Edgar's political contacts and Costa Rican media exposure would motivate the Costa Rican government to invite U.S. military support in the search. Meanwhile, Mead worked back channels to get special-ops soldiers, like the PJs, down to Costa Rica.

Mead, Josh, and Juan Edgar all thought it was a sure thing. Lauren was skeptical. The Costa Ricans pride themselves on having no military. "I'd love to see a Black Hawk helicopter land in Puerto Jiménez, but I'm telling you"—she smiled—"it's just not going to happen."

By now Roman's disappearance was national news. *Men's Journal*, ABC News, and a who's who of celebrities, media, and influential contacts from the Fellowship and Explorers Club all spoke to me. Midway through a marathon phone session we heard from Mead: "I just got off the phone with the number one motherfucker in charge of Southern Command—General Kelly!"

Four years before, four-star general John F. Kelly had lost his twenty-nine-year-old son—a marine killed by a land mine in Afghanistan. General Kelly himself had called Mead and told him he had learned just that day of the request to send special ops to Costa Rica. "I don't know what I have, or the legality," he told Mead, but General Kelly was on it.

I was honored, flattered, and a bit overwhelmed. I couldn't believe it. The cavalry was coming after all!

These powerful and successful people—media stars, generals, senators, governors—had strong but not overbearing personalities. They made things happen by applying vision coupled with persuasion. And they worked a network of social relationships that acted as "hands and feet" to accomplish shared goals. Josh and Mead (and the whole orb connected to them) were family and friend–oriented, less self-centered, and more likely to build and maintain relationships.

My own parents, grandparents, uncles, aunts, and cousins are good people, but my extended family seemed in tatters. I hadn't stayed in Virginia. My dad didn't stick around, my sister moved to London, and my own mom had left home when she was sixteen. They had their reasons—good ones. But mine were self-indulgent and selfish: head for Alaska to do what I wanted.

Peggy and I had worked hard to make our own family better than what we'd each grown up with. Our marriage, like many, had rough spots. Peggy once told me, "You haven't always been a good husband, but you are a good father."

My kids showed they loved me, even though I felt undeserving. Home from college, Roman once responded to my apology for being a less-than-ideal parent by saying, "No, Dad, you were—you *are* a great dad. I love you." My main hope was that Roman would be a better dad than I had been.

The next day, Alaska's Republican senator Lisa Murkowski emailed me with bad news. The Southern Command, air force, and joint chiefs of staff had concluded that the legal authority doesn't exist because the SAR isn't requested for humanitarian assistance or disaster relief. There'd be no American military-based search and rescue for Cody Roman Dial. The roller coaster plunged again.

During the full moon of August, I had looked up and imagined that Roman, too, saw the same bright disk in the sky: he was wondering where I was, when I would come get him. The image compelled me to return to Corcovado's mountains, jungles, and canyons. Fed up and exasperated with the effort to get Department of Defense support, I emailed Josh and Mead: I am done with this and have better ways to spend my time. Thank you two for your efforts. They were something to behold.

Lauren had been right. "The legal authority doesn't exist." I was disappointed that the cavalry wasn't coming. But not surprised. I couldn't help but think that the stink of the Pata Lora story had wafted its way north and suppressed any action. *If Jenkins's story is true,* and I believed that it was, *then where is my son?* It was time to go back to Zeledón.

Thanks to Juan Edgar's connections, Mead's endorsement, and Josh's push for media exposure, we had a mechanism to enter the park. And through Peggy's efforts back home, we had the services of three former military SERE (survival, escape, resistance, and evasion) experts from an Anchorage search-and-rescue training company called Learn to Return, or LTR. With our friends' support, we had the funds to fly them all to Costa Rica along with two of the Veracruz packrafting crew.

In his mid-fifties, Brian Horner, owner and founder of LTR, was skilled in search, wilderness medicine, technical rope work, and rescue. He had worked on projects around the world. Clint Homestead, in his late twenties, had served as a Green Beret in the Middle East and was skilled in rope work, too. Clint was muscular and fit and worked out at the same Anchorage gym as Jazz. The third crew member, Frank Marley, had been an army medic. Now in his thirties, I remembered Frank as a graduate student at APU.

Besides LTR, two friends that Roman and I had paddled with in Veracruz joined us: Brad Meiklejohn and Todd Tumolo, who'd led the way down the Big Banana. While Brad and Todd mostly packrafted whitewater with me, Todd was an accomplished climber and mountain guide in his mid-twenties. He'd helped me on some ice worm traverses, too. I'd met Todd when he was a student at APU, where he and Jazz had briefly dated. Like the LTR crew, Todd was trained in wilderness medicine. Brad, a climber and a skier in his younger days, spoke Spanish well. A professional conservationist and avid naturalist, Brad had visited tropical forests around the world. He had also become my primary whitewater packrafting partner in Alaska.

It felt good to have such a strong team of friends and community members ready and willing to head into the jungle. My only concern—as it had been with Ole and Steve—was everyone's safety. The afternoon rains were getting heavier, coming earlier in the day and sometimes lasting all night and into the next morning. The wet season had arrived.

THE MORNING AFTER returning from San José, a mysterious illness struck me with a pounding headache and dry heaves. After a near-delirious night, my sheets soaked from fevered sweats, I was just too sick to pack and plan. Josh and Vic cared for me. They brought fluids, food, and flu medicine from the farmacia in town.

I couldn't eat but the meds and fluids helped enough to get me out the door to the back-to-back meetings planned that day. Sick, my son missing for over a month, and suffering repeated bureaucratic roadblocks, I wondered, *Have my sins been so great as to deserve all this?*

Permission to enter Corcovado required that we fax twelve pages of permit applications to three offices. In addition, we presented our detailed plan at three meetings, complete with a day-by-day description of objectives, a list of our equipment, the qualifications of our team, and a communications plan. The permits would not come until the next day. All of this felt like inefficiencies in the system. But the hardest pill to swallow was that MINAE required that Dondee join us.

And there he was at a morning meeting, aching to be the center of attention, with his Google Earth projection of waypoints and GPS tracks on the wall. Dondee reminded us that Roman had entered the park illegally; he baldly stated that there was no place left to look in the park; it had all been checked already.

As he droned on describing odors from mining tunnels, and remembering the Tico guide's ballerina comment, I snapped. I'd had enough. This meeting was supposed to be about our plans for success, not Dondee's failure to find my son.

Standing up, I shouted, "We have been listening to this narcissist for a month and it gets us nowhere! I'm *tired* of it! *Fucking tired of it!*"

Dondee, satisfied that he'd finally pushed my button hard, smiled.

I stormed out to take a taxi back to the Iguana and leave him behind.

CHAPTER 35

Tree Fall

Steve Fassbinder on tree fall above Negritos Canyon, August 2014.

Six weeks after Roman walked into the jungle and a month after
I had arrived, MINAE finally granted me permission to enter
the park and lead a search of my own. Brad, Todd, and the three
LTR professionals joined me on Jenkins's route to Zeledón. Cruz
Roja, MINAE rangers, and Fuerza took a parallel tourist trail
and caught up to us later that day.

The crumpled landscape offered us few sites to pitch our
tents. Todd, Brad, and I set up a plastic Visqueen tarp and bug
tent camp where Ole, Steve, and I had camped before. The LTR

guys squeezed into a dome tent on another ridge along Jenkins's well-worn trail to the mining tunnels above the Negritos. Dondee, Cruz Roja, MINAE, and Fuerza camped near the north branch of El Tigre. One of the MINAE rangers was Kique, the tall, dark, and serious ranger who had hiked the Fila Matajambre ridge trail with Thai and Pancho the day we met Jenkins.

Jenkins had told me that on July 10, his brother, who had been with him and the other three miners on the Zeledón, had had a court date for his divorce. Walking downstream to make the appointment, Jenkins's brother had encountered Roman hiking upstream on El Tigre. This left several places to look for Roman between the Negritos and the north branch of the El Tigre. For example, there had been a rotten smell of decay at the mouth of Negritos's canyon. Looking there, I found a dead agouti, a spotted rabbit-sized rodent of the rainforest that looks like Borneo's mouse deer.

On rappel, the LTR team checked each side gulley leading into Negritos's canyon. The rest of us checked possible cliffs that Roman could have fallen from. It rained all afternoon and into the night. The next day we again looked hard, but all my ideas—the side gullies, the bad smell, the El Tigre's north branch—came up empty. These negative results reduced the number of places to look. While there was an infinite number of unlikely places—cliffs, thick bamboo, landslides, inaccessible canyons—there was only a finite number of likely ones.

Whenever I searched in the jungle, hope tugged me toward town, where someone might have found new clues or Roman may have finally revealed himself. But whenever I was in town, dealing with officials, reporters, logistics, family, friends, the cagey voice with the unlisted number, and all the rest, I just wanted to go back into Corcovado and look. Snakes, cliffs, rain be damned.

Soon after breakfast at dawn, Todd and I searched a landslide above the canyon rim. We walked and talked as we moved on and off trail. It was reassuring to have Todd along, a gentle, competent, intelligent young man. Todd said he'd always been a woods kid, and that his father had left his family to live in Panama when Todd was young. His story left me wondering as we came up empty of clues, if my son had left me to go to Panama. Maybe Dondee was right: he'd never entered Corcovado at all.

As the afternoon wore on and the short-billed pigeon called its melodious *who-cooks-for-you* from high in the canopy, we all rendezvoused back at camp. We'd found nothing.

THAT NIGHT IT rained hard and the wind blew. As snowstorms load mountain slopes that eventually avalanche, rainstorms weaken trees that eventually fall. But unlike avalanche safety, with its snow pits, locator beacons, and shovels to rescue an avalanche victim, there is no special technique, no technology, nothing like an avalanche-awareness class for safety from falling trees. Most people are surprised to learn that tree fall is a hazard at all, though the word *widowmaker* has been coined expressly for potentially lethal tree and limb fall.

In tropical rainforests, where trees are mostly shallowly rooted and where decay is rapid, thunderstorms can waterlog canopy deadwood with rain. The storm's winds can then break, snap, or tip up entire trees. Tim Laman once said about Gunung Palung: "The sound of tree fall is so common that I sometimes wonder why there are any trees left standing in the forest at all."

Brian, Clint, and Frank slept through the night's rainstorm in their small dome five minutes away from our camp. The wind picked up. Limbs started to fall, waking the three. But "with nowhere to go anyway," as Clint put it later, they just lay on their backs and listened, hoping for the best.

Sometime after three in the morning they heard a pulsing groan followed by an ominous *pop, pop, pop,* then an accelerating *swish.* Falling objects move fast, speeding up as they go. A blast of wind flattened their tent with a *woomp!* Then the tent popped back up and they found themselves unhurt, happy to be alive.

We paced out the length of the tree: 135 feet from root wad to tip-top. The crown's six-inch-diameter limbs had landed only ten paces from their small nylon tent. I shuddered to think what would have happened had it struck them in the night. It was another reminder of how dangerous the forest could be.

"There wasn't really much we could do," Clint joked the next morning, "except curl into the fetal position and mess our pants!" They laughed the tension-relieving laugh of battlefield soldiers.

BETWEEN US, WE had looked on the Zeledón's ridges and in its gullies; even farther afield when Brad joined Kique's mining camp raids. The thoroughness of our search left me 95 percent certain that Roman was not within a half mile of where Jenkins had seen him. But we hadn't thoroughly searched beyond that half mile.

Every doubling in distance from the point last seen tripled the additional area needed to look. No wonder the Cruz Roja gave up. The task seemed impossible. It was easier to accept that Roman had left the park and encountered foul play.

Or, like Todd's dad left him, Roman left us.

CHAPTER 36

Foul Play

Willim with dead fer-de-lance, Dos Brazos, August 2015.

The idea that Roman would one day resurface from a grand solo adventure—a possibility Lauren proposed in her cheerful voice—sure beat the alternatives. But the notion that he deserted us left me feeling that we had failed as a family. Dondee, Doña Berta, the sightings at Matapalo, and all the doubters who thought I saw only a son I wanted to see, had sowed the seed of his abandonment in my heart. But it wouldn't root. Roman wouldn't desert his family and friends. He was loyal to us all. He'd written a friend to say he looked forward to seeing her soon. He had recently exchanged instant messages maintaining a friendship that he'd had since grade school.

All the sights he'd seen, odors he'd suffered, and tastes he'd enjoyed on his tropical travels must have been full of family nostalgia for him. On our way back from hunting Tibetan ice worms in Bhutan's Himalaya, Roman and I wandered around Bangkok before flying home to Anchorage. It had been an incredible trip; we didn't want to leave. The stopover in Thailand offered us a last taste of the exotic. "This is how I remember all our trips to the tropics," Roman told me that night in Bangkok, "with us ending up in some big Asian city on your quest to find durian."

He found my stinky favorite fruit after first spotting a heap of mangosteens—his "yellow starburst with a tang" fruit from our first trip to Borneo—on a hawker's stand. "Look, they have durian!" he'd said, even though he didn't like the fruit any more at twenty-five than he had at eight when he'd written "worse than brustle sprouts! Yuck!" He sat with me on a city park bench and endured what he claimed smelled like garbage while I opened it up and ate it.

If Roman was not in the jungle and had not left for a new life, only the possibility of foul play remained, probably somewhere between the place he'd been last seen and Puerto Jiménez. Even Jenkins and his crew were not beyond suspicion. Kique suggested to Brad that we push for further investigation of the four miners as likely suspects. But I had spoken at length with them and followed behind Jenkins on jungle trails. My gut said they told the truth, that Jenkins was trustworthy. He had risked too much to be lying. I could not yet go down a path that seemed like betrayal.

The thought that a miner had walked Roman at machete point

up a side stream or hijacked him on a bumpy road was a parent's worst nightmare, but investigating foul play was beyond my experience. Sniffing out criminals would require experts.

Using Todd's satellite device, I texted from camp: Peg, he is not where I thought and pretty sure it must be foul play so I am coming home to you by end of this week. Talk tomorrow.

She texted back minutes later. Then we need to stay there.

Stay here? How come? Not any evidence.

WE ALL LEFT the jungle wet with sweat and shaken by tree fall. It hadn't rained the final night, so the creek was low, with good walking. Dondee drove us back to the Iguana. I thanked him in Spanish, shook his hand, even fist-bumped him. He had done his job. His guys liked him. And it looked like he was right after all: Roman wasn't in the park. Dondee drove off in the Cruz Roja Land Cruiser. We would never see or hear from each other again. LTR, Brad, and Todd headed home, too. They had offered just the kind of help Peggy and I needed and we were grateful. The Cruz Roja and MINAE were happy with them, too.

A few days later, the consul general and Barbara from the embassy drove seven hours to meet me at the lodge. The consul general, named Ravi, asked me things that went beyond the usual missing person questions. He wanted to know about Roman's equipment. Ravi didn't have much experience with outdoor gear, but he wanted to get it right. I pulled up Internet images of a folding sleeping pad and a Jetboil stove. These were items missing from the yellow bag at the Corners Hostel that I expected Roman would be carrying with him.

Ravi and I went through the possibilities. We agreed, given the searching, that it didn't appear Roman was in Corcovado.

After reading some of Roman's emails, Ravi also agreed he would not have deserted his family and friends. "That leaves foul play," I said.

"Or," the consul general suggested, "it's none of the above." He smiled. "Look, Roman, I want to assure you that even though the search is over, we at the embassy will keep the case open and coordinate further investigations."

Alone at the Iguana and emotionally spent after a month of searching, I wanted to leave but I couldn't. Not yet. Peggy and Jazz were on their way to Costa Rica, and Peggy had plans of her own. She needed to look in the park, to see how big it was, how hard it was. She needed, also, to find solace in searching.

There could never be any single moment—unless we held his bones in our hands—when Peggy and I could be 100 percent sure he was dead. But confronting the possibility of the profound loss of my son forced me to confront my own risk taking over nearly a lifetime of adventuring. After my close call on the southeast ridge of McGinnis Peak, I hadn't quit cold turkey, not really. There had been frozen waterfalls, whitewater rafting, tree climbing, glacier travel, and more—all of it risky, all of it thrilling.

For the first time, I realized how much suffering my death would cause in those who loved me. More shocking, though, was the fact that forty years had passed before I recognized this naked, obvious truth. The stark lesson masked by decades of selfishness was this: when I die, *I am dead*. I no longer feel anything. It's those I leave behind who feel the lasting pain: the more love, the more pain.

I didn't want to be the cause of their suffering.

CHAPTER 37

Peggy and Jazz

Peggy and Jazz at Iguana Lodge, September 2014.

Peggy and Jazz arrived on the Osa. It felt so good to have them near, to hear their voices, to see their cheerful smiles, to feel Peggy's warmth in bed and to touch her during the day. Having them present gave me far more than comfort. It gave me fuerte.

We tried to make it to Zeledón, but the rainy season had swollen El Tigre. Running brown and swift, the river turned us back half an hour upstream. Still, the day took on the feel of a tropical nature walk. We watched a small armadillo nosing around for termites and later saw a coatimundi, tropical cousin to the raccoon, climb a tall skinny tree. We marveled, like we always had, at the sights, sounds, and smells of the jungle.

We stopped to admire a poison dart frog mother, hopping along the forest floor with its tadpole clinging to her back, a miracle of motherhood in which she would climb into the rainforest's canopy to leave the tadpole in an epiphyte's reservoir. *Mothers are so tough.* I thought of Peggy holding up through all of this while I had been away.

Peggy and I had often considered Roman "mine" and Jazzy "hers" during their childhood. But Roman was every bit Peggy's son, too. We needed to work together on this, to rely on each other for complementary skills and temperament. Walking through the jungle, Peggy offered her thoughts. She wanted to put up new flyers with pictures of his gear: "To keep him alive," she said, "and fresh in people's minds. Someone knows something out there. They're just not talking."

Through our network of friends, a young American woman who had lived off and on in Puerto Jiménez contacted Peggy with a list of people and places on the Osa. "This woman told me we should go to another lodge—Danta Lodge. With helpful people and local trackers."

"Okay," I said, "we'll go there. Anything else? I'd like to hear your ideas. I'm running out."

"Well, I think we should walk the route Roman said he was going to do. Really check it out. Nobody has done it and we need to go and look there, where he said he was going."

We returned to the Iguana for lunch. The sea breeze blew the sound of gentle surf past palm trees and into the open Pearl. Roman, so clearly missing from what felt otherwise like a family vacation, obliged an unspoken agreement that we carry on as if we'd soon be sharing with him the delights we'd seen, then hearing his own humorous, self-effacing stories, his awkward guffaw.

As Roman's younger sister, Jazz had often felt overshadowed, but she needn't have. Even as a preschooler, she'd been the family spark, its nucleus. It was Jazzy whom Roman missed most on Umnak. She made him laugh; he took great pleasure in humoring her, too. Still, there was sibling rivalry, and while Roman was praised for his intellect, it was Jazz who had the highest grades, Jazz who had the common sense, Jazz who'd win at nearly every game. As much as I included Roman on outdoor adventures, Peggy did even more with Jazz at home and daily: making home-made Play-Doh, holiday cards, beaded jewelry, baking, sharing as only mothers and daughters can share. "I'm the only normal one in the family," she would remind us.

Before high school, Jazz discovered a variety of short-season summer camps: Super Camp, Surf Camp, and Golf Camp. She then researched, applied to, and attended each. She also wanted to play soccer, but unlike the shorter camps, and because we traveled in summer, there wasn't an opportunity for her to join a team. Instead, she became a competitive climber at the local rock gym in high school. She did well and competed at a national level. After graduating from Lewis and Clark College on a full scholarship and majoring in psychology, she took up body building and entered in a local competition, finishing fifth.

Ever since she was old enough to understand the chores assigned her, Jazz has been responsible and reliable. And since earning her driver's license, she has always held down a job. As a teen she shopped for groceries, drove the family car for oil changes and tire swaps. At sixteen, she sided our house with me, eventually taking the lead, thinking ahead, measuring then marking the boards for the cuts, and hitting the first nail. After college, she was the one who solved problems at home when Peggy and I were away, once calling the plumbers to unfreeze our frozen pipes after checking in at our empty house to find the water didn't work. Just like she had on the Harding Icefield, Jazz anticipates problems and asks the right questions to solve them.

But Jazz, I sensed, felt helpless in Costa Rica. As brother and sister, she and Roman had been very close, one of Peggy's goals as a mother. Being here likely distressed her, although she showed no sign of that. All of us had girded ourselves in search-and-rescue mode, but Jazz saw no reason to stay. She didn't want to do the hike Roman had sketched out and needed to get back to work in Anchorage.

After Jazz left, Lauren and Toby encouraged Peggy and me to assemble a poster offering a reward for Roman's missing gear. Todd Tumolo sent a photo of Roman's green Salomon shoes on his feet during our trip to Mexico. We copied and pasted Internet images of a yellow folding sleeping pad, a blue-colored Jetboil, a red dry bag, a puffy blue Patagonia pullover, the Kelty logo of his tent. I asked Lauren if we shouldn't hold back a few items, in case someone was—as with David Gimelfarb—trying to take advantage. "I don't think so," Lauren said, who'd been a defense attorney for a decade. "We want to get as much information out there as possible. It's time for a criminal investigation. Having lots of people looking for distinctive gear is useful."

We posted copies at every pulperia, *soda*, cantina, and colectivo stop between Los Patos and Carate. Everywhere we went, people sympathized with us, the parents of the muchacho. It felt good to be doing something new that might have results. For nine hours one day we hung posters, heard stories, peered into every backyard we drove past looking for his clothes on laundry lines. We studied every young man's feet for shoe color and brand, wondered about every kettle of vultures circling above forests and pastures.

One ex-pat said that when she had moved here in the nineties, it felt like paradise at the end of the world. Now, fifteen years on, she had nothing good to say. She wanted off the Osa but had her life savings tied up in her house and property. She had married a local Tico and had his child, but his substance abuse led to their divorce.

"I knew those two Austrians. I knew Kimberley, too, who was beaten and shot right at her house. And I knew Lisa, smothered in her bed. And you know what? I lock my front door each night, and then I take my child in my bedroom with me and I lock my bedroom door, too. I got a gun. I keep it loaded right there with me. And to top it all off, I let my ex live with me. Otherwise the criminals will just come and take everything. I'm trapped. Cody's disappearance is part of this," she concluded. "This inbred, lawless, uneducated, unscrupulous backwater of a place."

She stood up and took a deep breath. "Look, I'm always around. Stop by anytime. And sorry I just talked about myself. I hope you find your son. I can't imagine what a nightmare this is for you two."

CHAPTER 38

Cerro de Oro

Pulperia, Cerro de Oro, September 2014.

After my return from San José with Josh and Vic and ten days before Peggy arrived, Vargas had called me at the Iguana to say he'd heard a story about a lone gringo at Cerro de Oro, an off-the-grid mining community on the north side of Corcovado's mountains. Cerro de Oro is beyond La Tarde, where Dondee had abandoned Thai, Pancho, and me. To go there, Peggy and I hired a guide named Andres who spoke good English and knew the trails. Tall, young, and curly-haired, he led us to Cerro de Oro with the patient, attentive gait of a nature guide. He pointed out a mother sloth in a cecropia, a fast-growing, hollow-stemmed tree that looks like the house plant called an umbrella tree and is the sloth's favored food. Through our binoculars we could see the baby clinging to the mother's greenish-gray hair, looking down. Elsewhere, a stately king vulture, biggest in the Americas, dried its white wings at the top of a tall snag.

Walking upstream, we encountered a knot of miners studying their pans for gold. The youngest dug into the current with a spade. Andres told us a story about the oldest. With only two teeth and dark skin weathered from a life in the sun, he looked ancient as dirt. The old man had once been lost in the jungle for two weeks after breaking his leg. He'd been rescued when a local indigenous psychic described his location from a dream. One of the miners wrote the seer's number in my notebook. Every call rang busy.

The miners said there were two pulperias upstream. A three-hour walk off the grid, the first pulperia was crowded nonetheless. Under the rustic shop's roof, shaded by mango trees, a handful of men sat on wooden benches with dogs at their feet. A little girl peeked from behind her mother's skirt. On the store counter a green parrot cocked its head, eyeing us with the same cautious curiosity as the men, the dogs, and the little girl. "The owner says no gringo came through here," Andres told us. "You are the first ones here in memory." We hung a poster anyway.

A wide trail led to the next pulperia, ten minutes away through what was once a bustling village. Simple framed houses surrounded by gray fences stood empty in yards crowded by encroaching jungle. Cerro de Oro was a community reached only by foot or horseback, with running water gravity-fed through black plastic pipes from nearby streams. It felt forgotten by modern Costa Rica. "Roman wouldn't have even known about this place," Peggy concluded, "and I doubt he'd come here if he did."

I wasn't so sure.

At the second pulperia, a man and his young wife—or daughter, we couldn't tell—also reported that they had seen no gringos. The man said he had heard a gringo entered the park near the Rio Conte. The young woman shared some yellow rambutans. We thanked her for the tasty fruit, hung a poster, and left.

Farther along the trail we met an old miner. Andres asked if he had heard about the missing gringo. "Yes, of course," he said. "That muchacho has been lost before in the Amazon, and the father went down and found him that time, too." We had heard many rumors across the Osa, but this one was the most fanciful—so far. Later, I would hear even taller tales.

ON THE WAY back to the Iguana, we visited Vargas's farm to make arrangements for our trip to follow the route Roman had described in his last email, then stopped in Puerto Jiménez to eat Chinese food. Sitting in the restaurant, we watched a fight break out across the street at a liquor store. A couple of guys threw punches, rocks, and boards at each other. Nobody tried to stop them and the fight fizzled out on its own.

Watching this street brawl made it easy to understand why the Costa Rican government was closing down trails and requiring guides for all park visitors. The underbelly of the Osa grew by the year, people said, a place where convicted felons go to hide, where high-volume cocaine traffic flows freely from Panama, Colombia, and farther south. The miners, we'd been told, were drug addicts, self-serving thugs.

That night the heaviest rain of the burgeoning wet season hit with thunder and lightning that knocked out the Iguana's electricity. The wind blew hard from the Gulf. Tree fall crashed in the dark and I readied our things to escape should it feel unsafe on the second story of the Pearl, where we slept surrounded by tall ceiba forest.

By morning the Iguana's power was back on, the lodge intact. Lauren told us over breakfast that Vargas was nervous taking a woman along on our upcoming traverse. He thought Peggy would slow us and that he'd get caught. He said if we encountered any officials, he would run and wouldn't wait. He could go to prison for being in the park.

But Peggy, I knew, was much tougher than she looked. She had raced three times in the Wilderness Classic, holding the fastest female time for decades. She would have no problem keeping up. Lauren encouraged her: "Peggy, you need to go and show that old Tico what women can do. Straighten him out."

We went to town for cash. The sky was clear, the air wet, the sun cooking us overhead. My ATM card didn't work at the bank and I had too little Spanish to explain my problem. The bank declined my credit card. During a three-dollar-per-minute cell call to credit card services, I was transferred, put on hold, and asked the same questions repeatedly. My own went unanswered.

Frustrated, I vented on Peggy, telling her it was her turn to struggle with language, her turn to access money, her turn to drive everywhere. I would sit in the car and wait. Costa Ricans have a saying for misplaced anger like that: "I broke the dish, but you have to pay for it."

Jazz ultimately saved us, as she so often does. The bank said the easiest way to get cash was by MoneyGram. We texted Jazz in Anchorage. Within minutes she had transferred us the money we needed for the private investigator.

Emotional pain inevitably manifests itself physically, it seems. Leaving in the predawn darkness to meet Vargas at five, we hurried through the dark. Peggy couldn't fasten her seat belt because its ratchet caught with each jolt in the bumpy road. A cyclist appeared out of the black. I swerved, striking a deep pothole that sent Peggy flying out of her seat, where she hit her head on the roof, then landed on her tailbone that she had broken years before and bruised it severely.

She cried out, moaning in agony, tears in her eyes. I stopped, hurt by her suffering. I felt terrible, with no way to ease her pain other than with a gentle squeeze of her hand, a caress, an apology.

She motioned me onward. "Let's go. We're going to be late. I don't want to keep him waiting."

CHAPTER 39

Roman's Route

Upper Rio Claro, September 2014.

When we arrived at Vargas's farm in the dark on September 4 to finally walk Roman's planned route, it had been eight weeks since he wrote us his last email and forty-three days since we realized he was missing. We followed Vargas uphill in the tropical dawn, the best hour of every day, set between sleep and sweat. With his son Jefe in the rear, Vargas walked us along a dirt road, then an ATV trail, and finally a footpath where we slipped across the park boundary into Corcovado. Our shared language with Vargas would consist mostly of river names:

Agujas, Barrigones, Conte, Rincon, Sirena, Claro, El Tigre. We had no translator.

At first Vargas moved slowly and Peggy knew why. "He thinks I can't keep up. Tell him to go faster." She waved her hand forward from the wrist and frowned at Vargas.

"*Mas rapido!*" I said in my simple Spanish. Faster! When that didn't work, she pushed him with both hands and a smile. Physically urged onward, he looked at me quizzically but moved quicker with Peggy hot on his heels.

In the park, the trail narrowed and we squeezed past huge tree roots that sprawled across the narrow ridgeline like fat lazy pythons.

"Cerro de Oro," Vargas said, motioning to a side trail.

"La Tarde?" I asked, pointing ahead.

"No. Aqui," he replied pointing down the Cerro de Oro trail again. *Thai, Pancho, and Kique must have passed this way on their way to Dos Brazos.*

Dawn slipped away and by eight we were sweating. We passed beneath a gang of spider monkeys barking, screaming, and shaking the trees above us. We climbed higher. Just before mossy woods and overcast skies closed overhead, we took in a rare view of Corcovado Lagoon. Soon after we passed the Rincon benchmark and entered Las Quebraditas' disorienting bamboo forest.

At one of the *picas,* a small, subtle trail I would hardly notice, much less follow, Vargas led us to the Mueller benchmark. Startled, I turned slowly around trying to orient myself. For the second time on the summit plateau, I had lost my sense of direction. Peggy looked at my face. "What's wrong?" she asked.

"I thought we were going a completely different way and am totally turned around. I'm glad we're with Vargas."

"If *you* don't know where we are, then I'm glad we're with him, too!"

After climbing over deadfalls and past muddy trail braids, we came to where an empty package from Vargas's granola bar and heliconia leaves marked our lunch spot with Thai from a month ago. I looked at the litter and thought how many times I'd hoped to find a Starburst wrapper or other sure sign left by Roman. We sat on the same leaves on the same logs, knowing little more than the same facts about him we knew then. Peggy squirmed. Her bruised tailbone prevented her from sitting squarely on anything.

After lunch we moved down the broad ridge. Vargas motioned right: "*Sirena*." Then left: "*Rio Claro*." And behind us: "*Madrigal y Rincon*." We had passed the five-pointed star and now slipped through the keyhole that led off the plateau to Rio Claro.

At times, tangles of liana-choked deadfall pushed us off the ridge trail, but Vargas quickly got us on track. The gentle *tzing* of his razor-sharp machete left a wake of fresh-cut vegetation behind like bread crumbs. Smiling and joking in the oppressive heat and sweat-soaking humidity, Peggy watched for birds and monkeys. She never complained, despite waking up nauseated from dinner the night before and fretting about snakes underfoot today. She had no difficulty keeping up. This was the mother of our son.

By early afternoon Vargas announced: "Rio Claro." He pointed to a silver sliver far below at the bottom of a steep-sided valley. It seemed unlikely that Roman, following only the small, thin pica trails that are so rarely used and easily lost, would have made it to here. I hoped he had exercised judgment enough not to try crossing Las Quebraditas. But we had to. We owed it to ourselves and to him to be here and look. We kept close to the Osa's most experienced tracker.

On a ridge high above two branches of the Rio Claro, Vargas and his son disagreed over where we were. Once my phone's GPS acquired a signal through a thinning in the canopy, I showed Jefe our location and pointed to the Rio Claro on both map and landscape. Going right would take us down to the river, but instead Vargas took us left and then up, up, up, rushing headlong into a sudden downpour.

While trying to hang on to his pace, I gestured this didn't seem right. Vargas responded by plunging off the ridge on a tapir trail where the rainstorm left a small stream spilling down muddy steps. The pouring rain chilled Peggy first and then us all. "You said we'd camp before the rain! *That's what you said!*" Peggy reminded me over the din of big drops pounding on layer upon layers of forest leaves. "Why not *here*?" she implored.

"*Acampar aqui!*" I yelled over the crash of water. Camp here!

"*No agua!*" came the reply. No water!

I smiled and held out my arms, palms up at the deluge all around us, then dropped my pack, pulled out a large Visqueen sheet, and pitched it. Peggy and I ducked under the plastic to escape the cold rain. Vargas pulled out a brand-new, tiny dome tent and erected it quickly without Visqueen above.

Peggy collected rainwater in our bottles and cookpot as it ran off our plastic shelter. I erected our bug net tent, then took a full water bottle to the other tent as a peace offering. We stripped off wet clothes and hung them to drain. Peggy ate an entire hot meal, settling her cramps and relaxing her. While she'd found her appetite, she slept little, picking off small ticks from her skin as they bit her most of the night. Early in the morning, lightning flashed and thundered. A tree crashed to the ground.

FATHER AND SON sat eating breakfast as we broke camp and packed. My GPS showed us perched on a narrow ridge with the Rio Claro's headwater forks a thousand vertical feet below us. At first, we climbed higher, back into the dry-feeling oak forests, then dropped steeply down a knife-edged ridge to the Rio Claro. "It should be easier from here," I said, relieved to be down.

"I hope so. That last bit was too steep! And so much off-trail—I was worried about snakes."

At the bottom we caught our breath and scraped the leaves, twigs, spiders, and ants from between our shirt collars and sweaty, naked necks. "Roman probably realized this isn't a good route. It's more work than it is interesting," Peggy said between gulps of cool stream water.

At first the creek was shallow and slippery, but it soon gained flow from tributaries and its sandbars offered good walking. In quick succession, we passed an elaborate mining system of hand-built dikes, walls, and channels, then a beautiful natural weeping wall of seeps and waterfalls, covered in hanging gardens of ferns and mosses. The air smelled earthy and wet.

Peggy inspected areas along the creek that might hide Roman and his tent. "Where would he cross and where would he camp?" Peggy asked, tears in her eyes as she saw the immensity of the problem. "We need to get ourselves into his mind."

Vargas wanted only to charge downstream. Near three-thirty in the afternoon, Peggy said it felt like rain was coming. I tried to get Vargas to stop. Instead he shook his head and urged us on. We'd already suffered one icy cold shower. Peggy wanted to camp before another but it went dark before we could. Sharp lightning exploded in thunder overhead and the sky cracked open, dumping rain and chilling us instantly. We stopped, waiting for it to slow so we could set up the tents.

With no end to the deluge in sight, I strung a line and hung the Visqueen. We huddled under it, Peggy pushing us all close to her, body to body for warmth. The river rose past our toes, our ankles, our shins. *"Rio crescendo! Muy peligroso!"* said Jefe. River rising! Very dangerous!

In a flash, the Rio Claro flooded from a knee-deep, clearwater creek to a ten-foot deep, brown torrent, flooding the beach where we had planned to pitch camp. I left to find a campsite on an old river terrace in the forest, beyond the reach of rising floodwaters. By the time I hurried back to the others, the water was nearing their knees.

"Vargas, aqui!"

Together we wrapped plastic around a cross pole and resuspended our Visqueen on the old, forested river terrace. Peggy and I set up our bug net tent atop the palm and heliconia leaves that father and son had cut to cover the mud. We fell asleep to the roar of the flood. It seemed unlikely we would reach the beach the next day, much less catch the three o'clock colectivo in Carate, sixteen miles away.

In the morning, Vargas made it clear that he and his son were heading back the way we'd come. Sirena, hours downstream and a half mile from the mouth of the Rio Claro, had the highest concentration of tourists, guides, and rangers in the park. He couldn't risk being caught. Vargas gave us a machete, its point sharpened as a weapon. I shook his hand and then his son's. He gave Peggy a farewell embrace.

Startled, she looked at me and smiled big: "He just kissed me! *On the mouth!*"

The sixty-two-year-old grinned, turned away, and headed back upstream with his son in tow.

Peggy and I would be on our own.

CHAPTER 40

Rio Claro

Rio Claro, September 2014.

The heliconia and short palms were still wet with rain, but the bright sun shining in a cloudless blue sky left the jungle friendly again. Birds were singing and insects buzzing as another day got under way, as if the storm had never happened. The Rio Claro had crested hours before dawn. Logs and other fresh flotsam lined its banks, left by the receding flood. And while it was still high and brown, it was no longer pushy and we waded at times to our chests with little fear. Where the river narrowed between vertical walls that plunged into its waters making it too deep to wade, the reassuring *tzing* of the machete left a path clear of snakes, ants, and vines through the trackless forest. Peggy's bravery in the deep water and jungle amplified her beauty and strengthened my love and admiration for her.

Eventually the water dropped enough to reveal beaches with boot tracks. We followed the tracks to a trail, where we snuck along quietly. We were maybe twenty minutes from Sirena, the touristic heart of the park. Without the mandatory guide, we worried about being caught.

Walking along the maintained trail, we came to a puzzling arrow that a hiker had scratched across the path. Curious of its meaning, I turned and backed up in the direction the arrow aimed, nearly stepping on a fer-de-lance. A literal line in the sand not to cross, the arrow pointed directly at the snake in warning. The angry-faced serpent no doubt waited at the trail's edge for a fat rodent to pass. It would not have struck unless threatened—or stepped on. My foolishness elicited a nervous laugh from me and a head shake from Peggy.

More challenges awaited us at the coast. A high tide had pushed into the mouth of the Rio Claro. We also needed to dodge guides who might rat us out and rangers at the park boundary, ten miles away. It would be after dark when we reached the La Leona Ranger Station; we hoped to sneak by unnoticed.

My first attempt to cross the Rio Claro found me swimming and thinking about bull sharks and crocodiles. "I'm not swimming!" Peggy called out to me.

PEGGY WAS BORN the youngest of ten. She didn't learn how to swim until halfway through college. On our first trip to

Hawaii, when we were nineteen and twenty, she had not yet swum in ocean waves. Like many Alaskan-raised kids, she was afraid of deep water and could only "doggy-paddle," as she called it.

Somehow, I coaxed her into Waikiki's gentle surf. Heading into the shallow surf break, we held hands at first. But then, wading deeper, the surging ocean lifted her off her feet. Taut with exhilaration, she turned and wrapped herself around me. Her arms and legs felt warm in the cool Pacific as we drifted and bounded as one, my toes pushing off the bottom with each passing swell to keep our heads above water.

She clung to me, her smile wide as the sea. In that moment, I felt something for another person that I never had before: a *physical* sense of safeguarding and surety entwined with an emotional depth I longed to repeat.

It came again, just as richly with Cody Roman on a family camping trip to the island of Culebra near Puerto Rico. We had pedaled from the condo we rented in Luquillo to a nearby ferry dock, towing the kids in our bike trailer. Culebra is surrounded by coral reefs and we found a white sand beach where we camped in the shade.

A passing fisherman sold us spiny lobsters for a few dollars. Jazzy collected the foot-long seed pods of the *flamboyán* tree that she found fascinating, while Cody Roman, wearing his mask and holding his snorkel, bent over to peer underwater. I waded out to him and suggested we go deeper. The two of us set off to explore beyond the shallows where he normally stood.

At first, he rode on my back as I finned with my flippers. He clutched my neck with his left arm, holding his snorkel with his right. After he had ridden there a while, I reached to his hand with mine and he slipped off my back to glide along as we swam hand in hand.

It was a profound moment, somehow as deep as the instant I had witnessed Peggy give birth to him. There, over the splendor of a coral reef, to be so physically a part of his development, felt more enriching than holding his hand as he learned to walk.

The feeling intensified when we came to a deep channel in the reef and he climbed onto my back for the crossing. I could feel him tense up in my hand as the bottom fell away, then physically relax when on my back again. Once we'd crossed the channel, he slipped off to again swim hand in hand, but this time by his own volition. Without words he'd said, "Dad, I'm nervous and need reassurance." Then, "I'm okay, let's keep going."

Experiences like those with my family are moments I cherish. Enriching their lives with physical trust has always enriched mine. Roman and I would learn to scuba dive together when he was old enough and Peggy would learn more than the doggy paddle, but fortunately she wouldn't need to swim the Rio Claro.

MY FEET FELT a submerged sand bar that led all the way across the river. It would be no deeper than my navel. I came back for Peggy's hand and we carried our nearly empty packs on our heads as we waded across to follow a trail through the coastal forest. During the afternoon hike we saw a crested guan and a pair of great curassows, big, turkey-sized birds clumsily balancing in the trees. Later we watched a tapir feed on sour fruits with its dexterous, elephantine snout. Halfway to Carate, we stopped to watch the sun drop into the Pacific as the full moon rose in the east in a cosmic balance. With tide out and sun down, we slipped by the guard station unnoticed, limping on sore feet to Carate.

The next day we caught the morning colectivo to Puerto Jiménez, convinced Roman didn't complete the route he had described in his emails. "I think it's got to be foul play," Peggy said on the way into town. "He doesn't seem to be in the park. Maybe somebody's got him. Let's talk to the private investigator."

CHAPTER 41

Back to Alaska

Yaviza, Darién Gap, Panama, January 2015.

At the end of the first week of September, soon after we'd returned from the Rio Claro, Lauren put us in touch with Fernando Arguedas, the private investigator who had cracked Kimberly Blackwell's case and once been OIJ. Suspicious that people were telling us only what we wanted to hear, I gave Arguedas the names of those I'd interviewed to see if they would tell him what they had told me. He followed up other leads, too, talking to a dozen people altogether.

Most important, he asked Pata Lora our list of thirty-five

questions about Roman's gear, mannerisms, past, and intentions. The questions were meant to see if Pata Lora had actually been with Roman. Peggy even asked that Pata Lora draw Cody's tent and hairline and describe the shape of his hands. Her intuition is sensitive to people's behavior and motivations and her questions showed that perspective.

Arguedas and his partner went out each day to ask about our son, reporting their findings every night to Peggy, Lauren, and me at the Pearl. It was the rainy season, so guests were few. Arguedas spoke only Spanish, and I transcribed Lauren's translation into my notebook. At the end of his investigation Arguedas gave us a written report.

Among those who had seen Pata Lora and the gringo together, the middle-aged Arnoldo of Dos Brazos had the most details because the pair had stopped by and smoked marijuana with him at his house. Arnoldo told Arguedas on September 9, 2014:

I saw "Pata de Lora" walk by with a gringo and sit at the entrance of my house. They stopped to rest for a while and asked for water. "Pata de Lora" had marijuana in a plastic cup. It was around three ounces. "Pata de Lora" told me he was going to take him for free. The gringo told me they were going to "Carate." The gringo had money, food, a cell phone and a large camera. He was carrying a large backpack, blue. They were there for about half an hour, then packed their backpacks and left. Carate is about five hours away walking. The gringo was dressed with a blue-collar shirt. The gringo said they would return in eight days. The gringo had a good roll of money in a briefcase.

As with Arnoldo, most of the individuals interviewed by Arguedas confirmed and extended what we'd found already. Some were honest and said they didn't remember. Others recounted bizarre stories that didn't make sense. Pata Lora's answers seemed to blend fiction based on fact with fantasy. Nobody disputed Pata Lora had hiked with a gringo. The fantasy was calling him Cody.

During the week that Arguedas investigated, the OIJ brought a short-legged bloodhound to the Osa. Nose down sniffing for scent, the earnest dog's ears dragged in the jungle mud. An ad hoc group accompanied the dog and his handler into the forest as they headed for one of the tunnels where people reported the fetid smell of decay. A miner's bones would be found there two months later.

I had hoped to join the canine-led investigative team but couldn't, and instead ran to meet them on their return. I was overwhelmed to see Kique and Jenkins with the dog handler and Jorge Jimenez of the OIJ. Kique had convinced Brad Meiklejohn that Jenkins needed a more thorough investigation in his role in Roman's disappearance. But here were two sworn enemies— ranger and miner—working together, looking for my son.

The PI's report confirmed what we'd learned, leaving Peggy and me with no new direction to go. Maybe Roman had gone to Panama. Maybe he had been kidnapped, or worse. In any event, it was time to go home. We would return before Christmas to look in Panama. We said farewell to our friends, Lauren and Toby, and their sympathetic employees at the Iguana and flew back to Alaska.

ONCE HOME MID-SEPTEMBER, I felt emotionally drained, broken and empty. Peggy and I distracted ourselves with house projects and work. I had research reports due, classes to teach, graduate students to advise. Sitting in my office, my grad student Ganey came by to say that he knew that I loved Roman and that he was sorry Roman was missing. These simple words moved me, and I thanked him as he hurried, perhaps embarrassed, out of my office to work on his thesis.

Of course, I knew Roman loved me, too. I remembered the times he showed it, the moments he said it. Once, home from a two-month mountain bike trip the length of the Alaska Range, my hair wild, my beard long and thick, he said, "Dad, you look like what you are—an adventurer!" When he learned in school that the ancient Greeks espoused balance among mathematics, science, philosophy, reading, writing, and sports he complemented me in his understated way: "Hey, Pops, you would have made a good ancient Greek." But my love for him was obviously not enough. He was still missing.

Peggy and I watched escapist shows on Netflix in which investigators solved missing persons cases in a single episode. We binged on TV series that featured middle-aged and young men working together with mutual respect and a lot of good-natured back-and-forth teasing. Each night we would fall asleep to these diversions. My friends took me packrafting down my favorite run before freeze-up, then ice skating on wild ice across frozen lakes, rivers, and marshes.

On one long-distance skate trip in November, two of us flew to arctic Alaska. With backpacks and camping gear, we skated one hundred miles between remote Inupiaq villages in two days. Moving that fast so simply was exhilarating, invigorating, even momentous. In Kotzebue, where we ended the marathon skate, cell service was poor and I texted Peggy to tell her we'd made it and that I'd only fallen fifty times.

Well, I only fell once, she texted back, but I broke my wrist in three places!

Oh NO! I texted. Let me call you. I felt a familiar pang. Once again, I was guilty of being gone when a loved one suffered. I called Peggy on the hotel phone. Out at a friend's lakeside cabin, she'd caught an edge on her skates and gone down, reflexively catching herself with an outstretched arm but snapping her wrist instead. Lying there, she set her own broken limb, got in her car, then drove—alone and with one hand—an hour and a half to the hospital. She would require a surgeon to screw a plate to her arm bone, then a second surgery to have it removed once her wrist had healed.

HER ARM WAS still in a splint when we went to Costa Rica before Christmas. We drove around looking for green Salomon shoes on the wrong feet, familiar gear in a second-hand store, recognizable clothes hanging in a backyard. A Catholic priest took us to the small chapels that he served around the Osa. We posted flyers offering a $5,000 reward. Lauren suggested the amount as a believable figure and enough to motivate locals to look. It listed her and Jorge Jimenez's numbers.

Nobody called. Unlike David Gimelfarb, Roman was nowhere to be seen. Either no one was talking, nobody knew anything, or five grand wasn't enough to attract scam artists. Maybe he just wasn't there.

We drove to the Panamanian border a couple of hours south of the Osa. We met with the police, thinking maybe Roman had tried to sneak into Panama on his way to the Darién Gap, the last jungle on his checklist. We asked what happened to illegal entrants and learned that the police detain them until they have collected a sufficient number to send en masse back to their country of origin.

The embassy had inquired early on, but found no indication Roman ever crossed into neighboring Nicaragua or Panama or ended up in either country's hospitals or jails. Afterward, Peggy headed home and I flew to Panama City to rent a car and drive to the end of the Pan-American Highway at the Darién Gap. The half-dozen police checkpoints along the way question every driver and passenger in every vehicle. *Could Roman have possibly passed through all these without a passport stamp, like he had the Nicara-guan border?*

Yaviza, the village at the end of the road, felt hostile and dangerous with its grim-faced creoles, armed soldiers, bored Emberá natives, and end-of-the-roaders. After walking around the village on both sides of the river, hanging posters with the $5,000 reward, I spent the night in a guesthouse, shutting the louvers to keep out *Anopheles* mosquitoes from crawling into my room through the tattered screen windows. The room had no air-conditioning, no ceiling fan. Belly-up, sweaty, naked on a thin sheet over a soiled mattress alone in the dark, I reviewed the last five months. Our efforts, assumptions, and fears had crystallized in verse:

> *Trial and error,*
> *Failure and terror,*
> *The truth of the matter at hand.*
> *Death in a whisper*
> *Is so much to weather*
> *For the life of a wife*
> *And her man.*

CHAPTER 42

TIJAT

Carson Ulrich, Ken Fornier, Jeff Sells, Roman, and Peggy, Dos Brazos, July 2015.

The next day I drove back to Panama City, relieved to escape the Gap unscathed. My gut said Roman had never made it that far, but if he had decided to slip into Panama unannounced and undetected, he had succeeded.

By February 2015, we had run out of options. We decided it had to be foul play because no one had found any sign of him in Corcovado beyond Zeledón. This didn't mean that I wouldn't go back into the jungle to look, but it did mean we needed expertise that we simply didn't have: criminal investigative skills by an American, rather than—or together with—a Costa Rican. Ideally, it would be someone bilingual who knew how to get people to talk but who would also listen to us and learn what we knew of our son. It was a tall order.

For most parents of missing children, there is *no point*— until they are able to lay their hands on the remains of their offspring—when they will concede: *My missing child is dead.* Six weeks after he'd disappeared, the odds of finding Roman alive seemed even. But after six *months*, I knew enough biology and human survival to realize that the odds were nearly zero. Still, we had faith that he was alive somewhere, somehow.

A memoir entitled *The Cloud Garden* describes how its authors, Tom Hart Dyke and Paul Winder, were kidnapped and held hostage for nearly a year. Reading it gave us hope. So did the psychics who contacted us, performed a "remote viewing," and reported that Roman was still alive.

Over the following winter, television producers sought out our story, but we ignored them. We had been disappointed with media coverage of Roman in general. It had been sensationalized at best, exploitative at worst, and always mistaken in some way that heaped hurt upon our pain.

One television production company connected with Peggy through the Missing Americans Project. The project's founder, Jeff Dunsavage, maintains an online presence with updated postings of U.S. citizens who have disappeared while out of the country. Reading the monthly accounts of Americans, Canadians, and Europeans who disappear in Latin America is enough to give any tourist second thoughts about visiting there. Peggy found that the project's mission statement struck a chord and she joined. "Adding my son to the list," she wrote on the web page.

Dunsavage once claimed, "Media is the tail that wags the

government dog," and cynically pointed out that without the harsh light of publicity, public servants don't always serve. He emailed Peggy, then arranged a call with a television production company called TIJAT (This Is Just a Test).

A TIJAT producer told her about his own father, who had been murdered in Honduras. The producer spent a decade, he said, trying to get his father's murderer jailed, but without success. Then, within days of using a video camera as an investigative tool, justice was served and the killer convicted. The TIJAT producer found that using cameras opened people up in rural Central America in a way nothing else did. He suggested that TIJAT make a documentary film to speed up our search for answers.

I was doubtful about TV, but after her call with Dunsavage and TIJAT, Peggy told me, "They want to help and I think they can. Let's hear what they have to say."

TIJAT offered us a two-pronged effort to help us with permits and personnel. There would be a former Air Force PJ named Ken Fournier, who would help in the jungle, and a criminal investigator named Carson Ulrich. Short-statured, middle-aged, and muscle-bound, Ken and I knew each other from adventure racing and shared a mutual respect. Carson was a recently retired, twenty-five-year veteran of the U.S. Drug Enforcement Agency. With his bald head, goatee, tattoos, and towering stature, he looked like a guy who could kick ass and take names. These two were exactly what we wanted. The plan sounded ideal.

The producers said that they would step back and simply document the story as it unfolded. We were wary of reality TV. Mark Burnett's *Eco-Challenge* shows rarely looked like the adventure races that I had participated in, even when his camera crews followed and filmed my teams to feature. During our first conversation with TIJAT's producers, I asked how *they* differentiated between documentary and reality TV. After a long pause on the conference call, one of the producers volunteered that reality TV was "overproduced."

TIJAT would compensate Peggy and me for our time by hiring Ken and Carson. They would also pay us royalties for any family photos or videos they used. In June of 2015, we contractually agreed to this arrangement. It fit with our view that TV would keep the search alive, provide permits and expertise we lacked, and put pressure on Costa Rica's government and the embassy. But the arrangement came at an unexpectedly high price.

LOOKING BACK NOW and watching the resulting show, titled *Missing Dial*, I see what we gave up. We gave up the son we knew, the one we had raised, the one that I loved. We gave up Roman for a fictionalized character my voice-over called Cody. I read lines written by someone who knew neither Roman nor our history— lines I felt powerless to change and pressured to read. Worst of all were the dramatized scenes of our son's death, re-created to generate "buzz." TIJAT settled on National Geographic Channel as the network to fund *Missing Dial*'s production, in part because of my past history with *National Geographic* magazine.

Soon after we had signed the contract with TIJAT, the embassy let us know that they had possession of Roman's missing Mexican backpack. The big pack contained his sleeping bag, blue Kelty tent, Jetboil, and the cold weather clothing that he had used on volcanoes farther north. His pack also held his belt, an empty wallet, his blue jeans and flip-flops, new cotton socks still in the packaging, a puffy Patagonia pullover, notebooks, and more.

Half of what we had posted on the equipment flyer was there. In the photo I snapped of the yellow bag at Doña Berta's hostel the first day of my search in Costa Rica, the waist belt of the pack is visible in the corner of the frame. At the time I had no way of knowing it was Roman's.

Peggy and I were shocked that the embassy had held his pack for so many months before telling us. The OIJ had even received it from the new owners of the Corners Hostel months before that. Nobody had bothered to tell us until the final day of a *maximum* sixty-day holding period: they could have—but did not—tell us the day they took possession of it. Instead they waited months. Actions like this make the harsh reveal of public servants on television necessary.

Carson himself seemed to have a serious ax to grind with the State Department, while "production"—consisting of a constellation of a dozen producers and directors—had a stake in whipping up conflict for television drama. It felt to me as if both Carson and the primary face of production, executive producer Aengus James, provoked me to confront the embassy in its failure to tell Peggy and me about the very equipment I had described to the consul general nine months before. While I *was* angry about their failure, answers were more important: *What did he use as a backpack if he left both his Mexican pack and the yellow bag behind? And if not the Jetboil, then what stove did Jenkins see?*

Production put us up at an isolated eco-lodge on the Piedras Blancas arm of the Rio Tigre just past Dos Brazos. The lodge nestled intimately in the steamy jungle. Agoutis rustled boldly off the porch. A three-toed sloth climbed a short cecropia tree near enough for us to see the cloud of small moths that call its fur home. A rainbow flock of tanagers visited the bird feeders of ripe banana morning and night. Ken caught a fer-de-lance barehanded and brought it down to show us.

For Carson and Ken it was unbearably hot and humid, without electricity at night. As if to make us squirm and sweat even more during the jungle's daytime discomforts, production shined bright, hot, studio lights in our faces while Carson interviewed us. Straight-faced and sweating, Carson instructed me: "Tell me everything you know about Cody."

That would be impossible. Instead I recited what I'd told Dondee, and everyone else who would listen, the story, now old, about how Roman had been raised, what he'd done in El Petén, his disdain for guides and drugs. I told Carson about Jenkins and Pata Lora. I laid it all out. But Carson, like Dondee a year before him, didn't seem to be listening.

Meanwhile, Aengus wanted more emotion from me. "So the TV audience can better empathize," he said. He even staged Peggy in a scene along a jungle stream where she walked over a hazardous slimy rock again and again, in hopes, it struck me, that she might slip, fall, and grimace in pain, so the audience could better "empathize." I called him out. "A bit overproduced, don't you think?" I wouldn't stand for "reality" at Peggy's expense.

From the day production's team first entered the jungle, I wondered how a jungle search could have been part of their plan. Jenkins guided their team to Zeledón so they could film Peggy and me in the jungle with him. Half of the crew couldn't keep up on the trail; they lacked both fitness and experience. The sound man's shoes came apart. A cameraman slipped off the trail into a steep gully. We walked the last hours in the dark.

Production was unable to secure park permits. There would be no further searching inside Corcovado's jungle at all. Instead, *Missing Dial* would follow Carson Ulrich driving around the Osa, looking for evidence someone had murdered Cody Roman, doing what Peggy and I couldn't, what we needed Carson to do. And for that we were grateful.

Peggy and I had to head home to Alaska for some business, after which I would return to Costa Rica. On the airplane, we talked about how the show's production wasn't looking like it would become the documentary we were expecting. "But maybe when I get back down there it will have moved forward in the right direction."

Unfortunately, it hadn't. It moved back.

CHAPTER 43

Carson

Carson Ulrich, Iguana Lodge, August 2015.

I returned to Costa Rica alone. Emails from the show's producer Aengus, the director, and the producer's assistant promised important news: Carson and Ken are on fire. Good people are risking a lot to get us the answers. You're going to get the full download.

I wondered what it could be. It was obvious during the first week of the show's production that the OIJ was gun-shy of the media. The embassy, too, was unwilling to go on camera. And MINAE refused to permit park access to TIJAT. With neither park access nor my willingness to search fictionally for the camera, the production company focused on Carson and Ken working together to solve a murder on TV, something neither had done before, much less on TV.

The red-eye from Anchorage to San José in a seat that didn't recline left me spent. I slept twelve hours at the Iguana, recovering. When I finally emerged from my room, the crew was excited. Almost giddy, the show's director, Jeff Sells, who specializes in reality TV shows, led me to the Iguana's two-story postmodern hut and its upstairs dining room reserved for special occasions.

Sitting at the head of a table pushed aside for effect, Carson waited in shorts and a black T-shirt illustrating 100,000 years of weapons evolution. Ken sat at the table, too. Production put me between them. Three cameramen stood dripping in the noonday heat. Two shot from tripods. The third shot from his shoulder to best capture my emotions for an empathetic television audience.

"Honestly, this is not easy to tell you," Carson warned, leaning in. "Pata Lora took your son, and he met up with a group of miners. One of the miners' names was Guicho. They were using drugs."

He paused. "And they killed him."

This can't be.

CARSON'S BLUNT STORY was devastating. Its revelation of Roman's death was horrible to picture, but far worse was that this ex-DEA agent—who, I hoped, would find something new—had exposed nothing more than a sensationalized Pata Lora story.

"Why'd they kill him?" I asked, choking on the words both because they were so horrific to say and because I realized that Pata Lora was central to Carson's narrative.

"For whatever pocket change he had on him."

"They have a body?"

Carson shook his head. "This is the hardest part to tell you. They dismembered him." The image was horrible, even if I didn't believe it was true. "And they fed him to the sharks."

Carson delivered these statements as factually as if he had watched it himself, with as much empathy as if he were describing an oil change. He managed, "I'm sorry."

"How sure are you of this?" I had recovered from the bloody image as I grappled with the unthinkable.

"It's the only story we have and we're doing everything to corroborate it."

Frustration pushed my disappointment aside. We'd chosen Carson to do what we couldn't: investigate criminals. But here was the Pata Lora story again. I asked the obvious. "You don't have any other way to go, do you? I mean, that's it, right?"

"The only other alternative is they're all mistaken," Carson deadpanned.

Roman was raised in the tropics. He walked across the Petén, boated through the Moskitia. He planned to cross the Darién Gap. He would never take the tourist trail from Dos Brazos to Carate with a guy like Pata Lora. How can I make this clear? Why doesn't anybody listen? I know my son!

"Right, and that seems really unlikely, doesn't it?" I asked rhetorically, my blood pressure rising.

"It does."

"You guys got it figured out and all it comes down to is squeezing it out of him, right? There's only one lead and you got it. I guess it's solved. That's kind of how I feel." I was pissed off now. A year on and we were no further. In fact, we had slipped backward.

It felt like a wrap. I got up to leave. Aengus and Jeff had their showdown between stubborn father in denial and bully expert agent. They'd gotten me to choke on the words *why'd they kill him* while I was wired for sound and filmed. And the ex-DEA agent had his killer—Guicho.

But the whole encounter left me off-balance: the image of my son murdered, dismembered, fed to sharks, told coldly at midday under bright lights with cameras rolling to capture every twitch and tear for consumption and profit. This felt overproduced. This was no documentary. This was goddamned reality TV, and I had sold my soul for the wrong investigator.

That was harder to swallow than the dismemberment and the shark feeding. Carson's wholehearted conviction disappointed me. Here we were again, back to the Pata Lora story that wouldn't go away.

Maybe it won't go away because it's true?

WALKING DOWN THE stairs and back to my room I felt shocked, dizzy, out of body. The collision of *What if they are right?* with *Why the fuck won't they listen to me?* left me weak, shaky, vulnerable.

My phone rang as I wondered if maybe Pata Lora and this guy Guicho *had* killed Roman. The number was unlisted.

"Hello?"

"Roman." Even with the echo and delay of a cell-to-cell call bouncing off a continent's worth of towers, I recognized the cagey voice from a year ago, the one who'd asked if I had a weapon or someone to watch my back before I went in with Vargas to Las Quebraditas. "We have an asset in Costa right now if you need him."

"Huh? What's that?"

Maybe I do.

"I got one of my best guys down there right now. He's available and ready to take out that black snake. He could be on the Osa tomorrow." This was surreal. Here, I could take care of my next stage of grief—anger—with a simple *yes* on the phone.

But I was far from sure what had happened. And I certainly had no place for retribution or revenge.

I told the cagey voice no, we didn't need his asset, and hung up.

CHAPTER 44

Kool-Aid

Eyelash palm viper, Corcovado, 2014.

Back in my room, the ceiling fan wobbled. My confrontation with Carson and the caller's offer to take out the "black snake" had left me shaken. Maybe everybody was right, even the voice on the phone. Maybe I was just a father in a persistent state of denial, clinging to a romantic notion of his son.

Carson had nudged me away from my conviction that Roman had never been with Pata Lora. And while all of the Osa might have wanted to rid themselves of their pariah, everything I'd heard from officials was that Pata Lora had *no* record of truly violent crimes. As for this guy *Guicho*—perhaps *he* was capable of murder.

A week went by with me skulking around, scowling at Carson. Yet slowly but surely, Carson, Ken, and the production company and their ex-FBI-turned-consultants wore me down. Carson seemed hurt that I didn't accept his story as fact. Eventually, it dawned on me that you don't hire a consultant to argue. Carson had been hired for me. He was there to help.

Like me, Roman could be stubborn—glaringly, frustratingly, and passionately stubborn. I saw this in Carson, too, as if he channeled Roman's spirit, even his mannerisms: stiff hand gestures, a level gaze in distracted thought, a crooked finger. I empathized in Carson's frustration with me for neither accepting nor respecting his role. His inability to articulate his convictions, choosing forcefulness of expression over clarity of logic, felt familiar. Together, and perhaps ironically, these elements of Carson made me more favorable to his thinking.

Carson wanted the end game of an arrest. He was a cop, after all. But an arrest wasn't enough for me. Nor, at this point, was justice, even if Carson's story were true. And I certainly wasn't after revenge. Guicho or Pata Lora could be taken out, apparently, with a phone call. More than anything, Carson and production wanted to solve a murder on TV, but I needed all the pieces to fit without contradiction, without ignoring facts.

CURIOUS, I WENT back to Doña Berta and tested her memory of my visit from the year before. She remembered well Thai and me. She offered the same story: Roman had been there, left his things, paid to reserve a bed for his return, but never came back. She'd told Dondee something different, and Carson something else still.

For over a year, I had pondered each story a local told me in the context of directions given to strangers: even when locals have no idea which way to go, they give directions. It felt as if the locals on the Osa told us stories about our son the same way, but with a twist: they told us what they thought we wanted to hear. As if when asked, "This way?," they answered, "Yes," not knowing if it was the right way or not. No wonder Carson kept hearing Pata Lora stories. He paid his informants to tell them.

While many people feared Pata Lora, others simply disliked him. Some said that he had been imprisoned for murder. Pata Lora himself told Carson and Ken that he had killed a man over, of all things, a bicycle and gone to prison because of the crime. If that were true, then why did the OIJ and the Fiscal—the Costa Rican prosecutor involved in criminal investigations—both say Pata Lora had *never* been charged with murder?

It seemed to me that Pata Lora had psychological issues, not criminally violent ones. Firsthand complaints I had heard centered on his thieving and lying.

I certainly didn't have all the answers. I'd learned early on to get used to being wrong. But something was clearly missing.

If Jenkins had seen him cooking, how did the Jetboil stove get back to the Corners Hostel? Was there a second stove? And what was Roman carrying when he left the hostel? Was there a second backpack? The blue one that Cody carried with Pata Lora to Arnoldo's place in Dos Brazos and Roy Arias's house in Piedras Blancas? Or the green one that Jenkins had seen Roman with on Zeledón?

Carson had no place in his narrative for Jenkins and Roman meeting in the jungle and no use for days or dates, other than one Sunday in July when Pata Lora and Cody got in an alcoholic cabby's taxi (the colectivo doesn't run on Sunday). Carson completely ignored the previous year's account of the friendly guide with his distinctive ears who had seen Pata Lora with a gringo in Carate within a day of when Roman said he'd get out. *Anything* I offered Carson about Roman's character or experience was summarily dismissed as immaterial. Scientists call this kind of analysis "cherry-picking the data." Even Aengus, who'd hired Carson, observed: "Doesn't give you much faith in law enforcement, does it?"

Still, like Carson kept reminding me, a dozen people saw Pata Lora and Cody together. To bend the facts and fit Carson's story, I sketched out in my notebook the two jungle trips necessary during those weeks in July 2014 after Roman wrote his last emails and before Thai and I arrived.

On the first trip, leaving Puerto Jiménez soon after emailing us, Roman encounters Jenkins's brother hurrying downstream on July 9 or 10 for a court date July 10. Then Roman climbs Negritos's canyon walls, camps above Zeledón, and meets Jenkins the next morning, July 10 or 11. To fit Pata Lora's story with Jenkins's required that Roman walk out on July 11 or 12, leave the Jetboil and backpack at the hostel, hop in the drunk cabby's taxi with Pata Lora on Sunday, July 13, and walk to Carate by July 15, when Roger Munõz, the friendly guide, sees the pair. Then, sometime afterward, they meet Guicho, who kills, dismembers, and feeds Roman to the sharks. This way, all the dates would fit with people, places, and events. Now, I simply needed to change our son to someone we didn't raise.

Peggy reacted to my doubts with her own in an email:

> The women at the hostel need to be interrogated. They know something and need to talk. They are key. Nothing else makes sense. We, our friends, and his friends know our son, and know that he wouldn't even think of getting his hands on drugs—especially in another country.
> DON'T let anyone even try to sway you to join their uninformed opinion.
> WE know our boy, Roman. He would never be so stupid.

By the end of August, I'd been force-fed a narrative I believed. My journal recorded my feelings.

> We are closer than we've ever been to solving this and it's thanks to Ken and Carson. Carson says "they" want good TV. He says he wants a conviction, justice. He's in it for that. He also says this is for real and no TV show has ever done this for real in real time.

> *Carson must be right.*

CHAPTER 45

Pata Lora

Guichos, Carate, November 2015.

One of the biggest events in *Missing Dial*—when the producers and Carson are sure they've caught the lightning in a bottle of live crime-solving on TV—comes when Ken and Carson lure Pata Lora to a remote shack to scare him into thinking they have all the facts. Carson lays out the narrative his Dos Brazos informants had fed him, based on stories by Willim, who claimed his nephew Pata Lora told him about Guicho, the dismemberment, and the sharks.

Aengus put me in a stuffy SUV where I took notes, listening via an audio recciver connected to the car's speakers. A cameraman and GoPro recorded my reactions to Carson's soft interrogation of Pata Lora. At the end of a long hike together, Ken had led him to a remote shack where Carson was waiting. Initially wary, it took some coaxing on Carson's part and reassurance by Ken, Pata Lora's new best friend, to get him to relax and talk. That and chain-smoking cigarettes.

"Everybody saw you go in with him but you came out alone," Carson said referring to the hike with Cody.

"Who said that?" Pata Lora responded.

"Four miners in Carate." Carson's reference to the Guicho family was meant to throw Pata Lora off-balance.

Pata Lora now took the story and ran with it, spinning yarn faster than Carson could pick it up. He described in detail how he and Cody encountered the three Guichos a few miles up the Carate River from the beach. There was Pollo with a 9-mm pistol, Mario with a machete, and their dreadlocked father, who went by Guicho, standing behind his sons.

"I saw their faces, Guicho, like they were angry. They were looking at Cody, like, 'YOU OWE ME!'"

"What did Cody say?" Carson asked.

Pata Lora recounted Cody's alarm: "*What happened? What happened?*"

The Guichos responded: "Shut up, motherfucker—and you—you *run*! RUN! Or I will kill you, motherfucker!"

Pata Lora went on, "Of course, man, I run."

Carson pushed. "And Cody was alive?"

"Of course. I don't know how they killed him. I don't know nothing, man. I save my life, that's all. That was the last thing I saw. I'm not fucking lie. You look in my eyes."

The scene in the show is tense. The hidden cameras give it a peep-show quality. Pata Lora sounds shaken: "Now I'm fucking scared."

Carson was sure that this was Pata Lora's "dark secret," a true confession of a scene Pata Lora had witnessed—the Guichos abducting Cody.

The interrogation disturbed me, too. Afterward, everyone in production—cameramen, sound men, assistants, Aengus, Jeff—was quiet, somber, long-faced, and respectful toward me. Aengus walked up, his white iPhone earpiece dangling. "Sorry, man," he said putting his hand on my shoulder. I was in shock, feeling like I should believe what I had just heard, but couldn't.

Have I been wrong all along? And Dondee and Carson right? Was my son not who I thought he was?

CARSON CLAIMED "THIRTY witnesses" from Puerto Jiménez to Carate saw Cody and Pata Lora. Whenever I challenged him, he responded, "They're all wrong, and you're right?" Dondee had said the same thing. It was what everyone said, the polite ones, too, when I was out of earshot: "That poor grieving father. He's lost his son and refuses to accept the truth that his son had no skills and poor judgment."

For me, the show *Missing Dial* doesn't document Carson and Ken catching a killer. It documents my betrayal of the son I raised as I warp dates to fit the fantasy of the Osa's pariah: Pata Lora.

I managed to save some of my son's dignity in a corner of my heart, crowded against our Umnak walk, Culebra swim, Wilderness Classic, and a lifetime of trust. I tried to make sense of these feelings in my notebook:

I'm wondering how did I just let everything I know go—I was so sure that Roman wasn't with Pata Lora. Now I've given in to all of it, following Carson's lead. Like him, any little thing that doesn't jive with the narrative is just <u>ignored.</u>

Dates don't line up?
Nobody keeps a calendar here.

Footwear and colors don't match up?
What color did you wear last week, last month, last year? How about just yesterday?

Smoking pot?
Are you so sure it wasn't just Pata Lora smoking pot?

Thumbtacks?
Maybe he was trying something out for Panama.

And a guide?
Maybe he just liked Pata Lora as a local, was using him the way he might in Panama.

Just writing all that is difficult! It's like I feel like I am stepping on his memory! Like I am disregarding everything I know after six months of emails and a lifetime with my own son.

But I signed on for Carson and this is where he brought us—and while I have trouble separating his truth from his manipulations, he does want to solve this. And he has "dozens" of people who said, Yes, we saw Pata Lora with that gringo who looks and acts like Roman! That seems less likely than the other contradictions.

Do I have doubt?
Yes.

Am I willing to swallow that bitter pill?
Yes.

Do I want this to end?

Of course.

By September 2015, I had stepped out of everyone's way, tried hard not to disagree anymore, and let Carson and production work with the retired-FBI-agents-turned-Hollywood-consultants to move forward. The consultants, Carson, and production all agreed that Pata Lora's recorded statement was sufficient for an arrest and a conviction of murder.

AFTER CARSON'S INTERROGATION, I volunteered to meet with Pata Lora. Aengus and Jeff could hardly hide their excitement: maybe as the father I could convince Pata Lora to go to the OIJ with his story about the Guichos. Six cameras recorded our meeting at the waterfront in Puerto Jiménez. It was in front of the seafood restaurant, where Peggy and I had spoken on the phone the night we last felt that Roman was all right after all.

Pata Lora had always been a sideshow for me, someone I didn't need to talk to—the personification of a rumor I never believed. I had discounted his stories as fabrications, a grab for attention, tales that rubbed the salt of insult—dishonoring Roman's skill and independence—on my raw wounds of loss.

This would be the first time that we would meet. Carson and Ken had showered him with gifts, meals, hotel rooms, even cash. He arrived in their car with them.

After production wired Pata Lora for sound, he walked up in sunglasses to meet me. Lean and about my height, he wore a green camo-colored ball cap, board shorts, and an army-green shirt that Carson would wear. Other than eyebrows and a thin, well-trimmed goatee and line of beard along his jawline, he had no hair showing on his head. He walked with a determined gait and a barely perceptible limp. His left ankle bulged in deformity. It looked like a parrot's foot.

I asked Pata Lora to tell me the truth, to make nothing up, and to take off his sunglasses. I wanted to see his dark eyes as he shared vignettes of my son's final days. Instead, Pata Lora gave a coarse description of a hike with Cody that left me unconvinced he had ever been with Roman. There were no details, no stories, no images. Looking into Pata Lora's gaze I saw only emptiness.

His eyes came alive when he described his own abusive father who had abandoned him to grow up on the street. Pata Lora acknowledged that he admired me as a father looking for his son. At the end of the made-for-reality-TV moment, he asked for a hug. I gave him one, feeling sorry for Pata Lora, but unable to shake the feeling he had never been with Roman.

Between takes for *Missing Dial*, Carson led a crusade criticizing the OIJ and the State Department for their inaction. It was hard not to admire his passion, but it got him into trouble. During one meeting with the Fiscal, Carson inadvertently wore a hidden wire. The government threatened arrest with a *denuncia*, a formal legal complaint or warrant. Carson fled Costa Rica, his mission to solve a murder in real-time TV unfulfilled.

With Carson gone, it was now up to me to carry on the new Pata Lora story during multiple meetings from Golfito to D.C., often with Aengus and Ken in attendance. Flipping through a folder of eight-by-ten-inch photos of a dozen witnesses from Puerto Jiménez to Carate, I told Carson's latest version of the Pata Lora story to the Fiscal, the embassy, the director of the OIJ, and eventually, with embellishments of my own, to the assistant director of the FBI.

I had to admit—as awful as the sweaty filming and bullying by Carson had been—the heavy presence of *Missing Dial* accomplished what Mead Treadwell, the Fellowship, GoFundMe, scores of volunteers, and Facebook posts had not. It had put sustained pressure on everyone. By September 2015, everybody from Osa's illegal miners to the director of the OIJ had one thing in common with Peggy and me: we all wanted the search to end.

Missing Dial had everyone's attention and it hadn't even been produced. The officials probably feared how their actions—or inaction—would look on TV when the show finally aired. Carson and Ken, it seemed, got more people talking in a month than OIJ or the FBI had in a year. Still, according to Costa Rican law, a body is needed for a murder conviction. Without that, Pata Lora's recorded statement during the interrogation, while valuable to the Fiscal, was not enough. Pata Lora's story might just be a confusion of fact and fantasy.

The day I left Costa Rica, Pata Lora took the boat to Golfito, promising to give the Fiscal his statement identifying the Guichos as having abducted Roman at gunpoint.

CHAPTER 46

A Backpack

Mall, San José, Costa Rica, March 2016.

I flew home to Alaska the first week of September 2015 and went back to work teaching a full load of classes, writing papers, advising students. We had hired a lawyer who petitioned the court for guardianship of Roman and in November we successfully subpoenaed the bank for his 2014 financial records. The records showed what we'd known all along: after July 9, the day he wrote us, there had been no bank activity.

Through the fall and into the winter, Peggy and I clung to Carson's embellished story. It was all that *Missing Dial* could give us. The price I had paid for drinking Carson's Kool-Aid and telling an enriched Pata Lora story was high: denying the life I'd known as a father to my son, as if our lives together had never happened. But the benefit was that Costa Rican and American officials believed the story.

All that OIJ needed was a body between Piedras Blancas and the Pacific Ocean for corroboration, conviction, and justice. And if that happened, we'd have closure, albeit incomplete. We'd know what happened physically—who killed him and how, perhaps—but we would never know why Roman became someone who would walk with Pata Lora in the first place. And that question nagged me.

In January 2016, the Fiscal separately informed both Aengus and me that psychological testing of Pata Lora diagnosed him with schizophrenia, a condition that explained much of his behavior. About the same time, Peggy and I finally had what we'd always hoped for: park access. Accompanied by embassy officials, Fuerza police, MINAE rangers, Cruz Roja (without Dondee), and the OIJ detectives and dog teams, we could go anywhere we wanted—just so long as TIJAT wasn't there.

FROM JANUARY THROUGH May, Peggy and I made four trips with these Tico teams. Sadly, each trip was a search for Roman's remains rather than his broken, living body. The search teams that worked with me now had come two years too late. This was the kind of support I had hoped for in 2014, when he had possibly been alive, when I had wanted—but was denied under threat of arrest—access to the park. The leader of these searches was a Tico lawyer named Jorge who worked at the U.S. Embassy and whose father once directed the OIJ.

On the first trip, Jorge picked up Peggy and me at the airport and drove us through the busy streets of San José, explaining the Costa Rican judicial system. "Mr. Roman," he said, "in Costa Rica it is essentially impossible to get a murder conviction without a body. Unlike in the USA, people saying things is not enough. In fact, the murderer could even confess to a killing, but without physical evidence, like a body, there could be no conviction."

Jorge had passed tests to become both a Fiscal and a judge and knew well what was necessary for justice. "Without a Fiscal and a judge present, OIJ investigators cannot even ask any questions, other than where a suspect lives, his name, and other nonincriminating information. All of this makes this case with Pata Lora very challenging."

With Jorge, OIJ, its cadaver-sniffing bloodhounds, MINAE, and Fuerza we searched between Carate and Piedras Blancas. Local miners helped us look off-trail and in mining tunnels that honeycombed canyon walls. Ever since my first days in the jungle, I had made a habit of looking among the miners' few possessions under their open black tarps. And here, on the banks of a small creek, I spotted something familiar beneath one. It was a short piece of foam sleeping pad of the type I recalled giving Roman two years before in Veracruz.

I had packed our packraft paddles with small pieces of pad like this on my flight to Veracruz, then offered it to him as a useful piece of gear. Its color, type, brand, and dimensions matched a pad I had once trimmed for an adventure race a decade before. It was the only physical evidence I had ever seen of Roman in the jungle. And there it was on the Pata Lora trail.

Questions flooded my imagination. *How did it get here? Are Roman's remains nearby? Is this miner involved?* The OIJ and Fuerza swarmed over the old man who was just downstream with his gold pan. The miner explained that he had bought the pad in a community near Dos Brazos years before. Suspiciously, he also lived with the woman who'd raised Pata Lora after Pata Lora's own parents had estranged him.

IN MARCH 2016 during Anchorage School District's spring break, Peggy and I again headed to Costa Rica. We spent a few days in San José where I hoped to discover what Roman had purchased for the $411.91 his bank records showed he had spent there. Throughout Latin America, his total monthly expenditures had generally averaged about $1,500. This purchase was a significant outlier. Sitting at the desk of our airport hotel room, I studied two lines on the bank statement.

07-06 WITHDRAWL DEBIT CARD PURCHASE
   $411.91
07/05 PURCH 2438921418641877318698 TNF 04 SAN JOSE
   CR

Googling 2438921418641877318698 TNF 04 hit nothing. I puzzled over the three letters *TNF*. *What's TNF?* The family tent that we'd used for years and pitched on Kuyuktuvuk Creek and Umnak left me thinking, *TNF. . . . Could TNF stand for "The North Face"?*

I Googled San Jose North Face. A store nearby in a large shopping mall popped up.

"Peggy! Maybe Roman bought a GPS at a North Face store here in San José. People who saw him with Pata Lora mentioned a GPS and a camera."

"It's a lot of money," she said. "Maybe he bought a camera, too. And shoes? Raingear? For his Darién trip?"

Peggy and I jumped in our rental car and hurried to the mall, excited that this might answer some questions. Soon the store clerk was paging through cash register records for July 2014. At nearly 220,000 colónes, Roman's purchase was easy to spot on July 5.

But he had not bought a GPS or a camera—the store didn't sell either—it had been a *backpack!* Midsized and lightweight, the Conness 55 model North Face backpack was well suited to the style of wilderness trekking that Roman would do in Corcovado or the Darién.

This is the missing pack!

I photographed pictures of the pack from their catalog. We now had a new search image while walking in the jungle: an olive-gray, midsized pack with a zippered compartment on the bottom and pockets on the waist belt. Excited, I texted Aengus about this important news. He texted back with the kind of lukewarm response Dondee had shown when Thai and I had found the yellow bag at the Corners Hostel; probably, I thought, because it didn't fit his show's storyline.

We had also heard a new spin to the Pata Lora story. An Osa Tico told us in November 2015 that Pata Lora *twice* confessed to him that he—Pata Lora, not the Guichos—had killed Cody, then buried the body. The Tico suggested that we offer a six-figure reward, like $100,000 or more, to lure Pata Lora into revealing where he'd hidden Cody's body. The local said he could facilitate. Using only the *promise* of the money, together with drink and mota, the Tico said he could convince Pata Lora to reveal where he'd hidden the body. We would pay nothing. As we drove the seven hours to Puerto Jiménez from San Jose, Peggy and I discussed how we might trap Pata Lora.

Our third day in Costa Rica, March 2016, Peggy and I drove to Pata Lora's house. We hung out with him and his French girlfriend, had a beer, made small talk. Pata Lora rolled a joint and shared it with the mademoiselle he lived with and who paid rent on the house.

As we were leaving, I pulled out a flyer offering a $50,000 reward—ten times what we had posted more than a year earlier for Cody Roman's remains. I had given another to the local Tico whose idea it had been in the first place, encouraging him to follow through with his plan. We hung a third at the Dos Brazos pulperia. All were meant to flush out the killer.

Pata Lora hated his nickname, so Carson and Ken had taken to calling him "Joe," an Americanized version of his first name, José. "Listen, Joe," I said, handing him the flyer, "we are offering a big reward to find our son's remains. Fifty thousand dollars." I looked to see if he'd take the bait, but his face didn't change. He had the same nearly blank expression he showed when I asked him to recount his days walking with Cody. "Maybe, since you saw him with the Guichos, you can find him?"

"Yeah, I can, man. *Sure I can!*" he enthused. "But I need some scuba diving equipment, so I can go into the ocean out by Madrigal River where his bones are. His bones are underwater, man," he rambled on. "Can you give me some money?" He took another toke off his joint, then offered it to me. "So I can buy the equipment and go into the water and find the bones?"

I declined the smoke and finished my warm beer. "No, Joe. We'll give you money for the body when you find it. Maybe you can get someone else to loan you the equipment and you can split the reward with them."

THE NEXT DAY, Peggy and I headed to Piedras Blancas in a black SUV with an embassy driver. Jorge, two OIJ detectives, and a pair of cadaver dogs and their handlers drove in their own vehicles. We would stay at Roy Arias's place, then hike on and off-trail to the ocean by way of the Carate and Madrigal Rivers inside Corcovado. The SUV climbed a greasy jeep trail through deep ruts and along narrow ridgelines leading to Piedras Blancas.

Peggy was nervous with the driving conditions. At one point, a stick jammed in the undercarriage. The driver stopped and Peggy jumped out to remove it. The SUV was perched at the edge of a steep cliff with a few trees near its top and an erosion ditch that cut into the jeep track. After Peggy hopped in, the driver pulled forward. The ditch grabbed the front wheel, pulled the vehicle off the road, and rolled us down the hill.

In the roll, Peggy, who hadn't had time to buckle her seat belt, was thrown forward from the back seat behind the driver to the front seat's passenger-side window. When the SUV came to rest, held in place by trees on the brink of a precipice, she peered, terrified, down the cliff below her. Unhurt but shaken, we climbed out the windows and scrambled back to the road. Squeezing in with Jorge, we tried not to look out the window at the roadside cliffs for the rest of the drive.

We slept on Roy Arias's hardwood floors, leaving at dawn for the Madrigal. Following a thin trail along a canyon rim, Peggy was stung by a six-inch caterpillar with long poisonous hairs. Later, bushwhacking a route we suspected the Guichos used to access the Madrigal River inside the park, we nearly stepped on a coiled fer-de-lance. Slowed by steep climbs, slippery descents, and sketchy traverses, we didn't make it to the main branch of the Madrigal until dusk.

Among the seven of us, we carried only two flashlights. Walking down the Madrigal in the dark, Jorge instructed us to walk through the creek's waters "to avoid stepping on snakes attracted to the stream by frogs." The warning seemed silly and walking in the creek was difficult, so Peggy and I climbed up to walk on a gravel bar. Not ten feet later, she stepped on the rubbery cordage of a snake. We didn't look down to see what kind it was or what it did. We just rushed for the creek and splashed along with the Ticos.

We reached Carate near midnight, finding nothing but bad luck and near misses between Dos Brazos and the Madrigal on the Pata Lora trail.

CHAPTER 47

Discovery

El Doctor Creek, Corcovado National Park, May 21, 2016.

By the first week of May 2016, it was clear that the series *Missing Dial* had been produced for the National Geographic Channel as reality TV. The trailers were ghastly. Their reenactments of the Pata Lora story focused on blood dripping from a machete. The machete was held by a man in board shorts and knee-high rubber boots standing over a body facedown in a creek. The show's executive producer, Aengus, even cautioned Peggy and me not to watch. He said he didn't want these re-creations, "but the network asked for them." He had produced them to create "buzz."

At the time I believed him, and I believed Pata Lora, and I accepted that the overdramatizations—which included a young man who looked remarkably like Roman in the photo from Bhutan but running from miners with machetes—would keep the investigation alive. After Aengus's warning, we watched the six episodes of *Missing Dial* that TIJAT had so far produced. I emailed him:

> Peggy and I watched all the episodes.
>
> It documents the investigation well.
>
> It's not overproduced in general although as you warned the machete scenes are a bit overused. My mother and sister would be disturbed by those. Probably Jazz too.
>
> It doesn't make the Embassy look bad at all, so if you gave them the episodes then they'd have little incentive to do any more than they've been doing—which has been very little really.
>
> I like the idea of using this show to leverage some action from the Embassy and OIJ.

As promotion for the show, a TV camera followed Peggy, Ken, and me as we headed to the FBI headquarters in Washington, D.C., where Ken had arranged a meeting. It was Thursday, May 19, and the next day we were scheduled to fly to New York for another promotional spot. The first episode would air on Sunday.

We left our cell phones at security. Two friendly FBI agents in dark suits led us to a small conference room on the seventh floor. The room quickly filled with seven or eight more agents, including the deputy director. I told the story I had come to believe, the one carved from Pata Lora's schizophrenic psyche, embellished with the Guichos and a new guy named Poquito, who seemed to be the Guichos' boss. I also told the agents how I had tracked down Roman's backpack purchase to the North Face store in San José.

"That's real investigative work," an agent said about the pack. "It's the kind of thing that we do. But what do you guys want?"

I answered his question literally because I'd learned over the years that the FBI couldn't really do much in Costa Rica. "For a long time, I just wanted to know what happened. Now we know. Now I want justice."

"A body," Peggy blurted. "We want a body. We want to bring him home."

A body would answer a lot of questions. It was also necessary for any murder convictions.

The rest of the meeting was like so many others over the years: assurances and explanations about the limits of American law enforcement in foreign lands. "These things take time," said the agent who had flattered me. "It could be years." They also commented on how Carson's activities had damaged relations between the U.S. and Costa Rica on this case.

AFTER THE MEETING, we collected our phones. Mine had a message from a number in Costa Rica. It was the consul general from the U.S. Embassy. He said to give him a call, no matter the time. My phone was almost dead. "It's Ravi. He wants me to call," I told Peggy. We walked the few blocks to the hotel to charge my phone. I called Ravi and put him on speaker.

"Roman," Ravi said over the phone, "I'm not sure there's any other way to say this but directly: human remains were found today near Dos Brazos. With camping equipment."

I sat down. Over the years, Toby, Lauren, and the embassy had contacted us about news of other bodies in the jungle. But this felt different. This felt like Roman.

Ravi continued. "What we understand is that a miner had been in the mountains today and found bones in a streambed. Then, moving upstream, he found camping equipment. He immediately called 911 from there in the jungle. We wanted to let you know as soon as we could. It seems this might be your son."

My feelings swirled between pain and relief. Relief, because it seemed the ordeal of searching without knowing might be over. Pain, because it would mean, once and for all, that our son was dead.

We needed to return to Costa Rica immediately. I had to see the scene, to judge for myself if it had been crime or accident.

Two years before, I had described in detail Roman's equipment to Ravi. The blue Patagonia Puffball. The Jetboil. Green Salomon shoes. A yellow and gray Z-rest foam pad that folded rather than rolled. Peggy and I had made a poster of these and others items, hanging copies from Cerro de Oro to Carate. But I hadn't listed all the things I knew Roman carried. I kept some to myself, for later, for proof. For a moment like now.

"Where did they find him, Ravi?"

"Up the Rio Tigre from Dos Brazos. Inside the park, in Corcovado. MINAE rangers are going up there in the morning to confirm."

The next morning, Friday, Peggy and I went to my mom's, taking the train to northern Virginia where we waited for her to pick us up at a bus stop. While we waited, a writer from *People* magazine called to interview me for an article about *Missing Dial*.

In the middle of her questioning, my phone lit up with a Costa Rican number. I told the reporter I had to take it. It was Kara, from the embassy. She was brief. She said she'd send me photos from the site where the remains had been found. She asked that I confirm if I recognized any equipment.

The photos arrived on my phone and I hurried through them. I needed to know if this was him. The first showed a bright green Salomon shoe pushed against a fallen tree limb, toe down, half buried in sand and debris. It looked as fresh as if it had come right off Roman's foot that day. "His shoe doesn't look old enough to have been in the jungle for two years!" Peggy exclaimed.

The next photo showed a pack, bottom up and partially beneath a rotten log. It, too, was mostly buried in dirty sand and gravel among sticks and brown leaves that had obviously been washed down as flood debris. A cookpot was partially exposed next to the pack. Another photo showed the pack free from the log and debris. It looked greenish gray. I quickly pulled up the catalog photo from the San José North Face store and found myself catching my breath. It was the same model and color.

The shoe and the pack were convincing enough to answer Kara's question. But as more photos arrived, there could be no doubt. They were all Roman's things. The yellow and gray colored folding sleeping pad was crumpled and partially shoved beneath a log. A compass with a black lanyard. A blue Petzl headlamp that I'd handed him in Alaska. This was all essential tropical camping equipment from our family stash back home.

There was also an unfamiliar silver cookpot and something green and metal. I couldn't judge its size. It rested in a streambed where shallow water ran over it. I didn't recognize this object at all. The last picture was sobering. It felt callous that they had even sent it. It was unmistakable: a human skull with the upper jaw visible, half buried in sticks and debris and backed up against a termite mound. Everything looked to me naturally deposited by flowing water in a creek, not haphazardly buried by a criminal hiding evidence.

My battery was nearly spent. The reporter from *People* called back. I sputtered something about the discovery, but her interview seemed superfluous, and I said I had to go. Peggy and I studied the images, teary-eyed and nostalgic. We reminisced about Roman, held each other and cried on the bus stop bench.

We pored over every image again and again, checking and re-checking that these were indeed his shoes, his pad, his pack, his headlamp, his compass. There was no doubt.

After the miner had dialed 911 from one of Thai's "little jungle phone booths," the news traveled fast. Texts and emails poured in from Lauren and Toby soon after those from the embassy. The Cleavers knew the ranger who had gone with the miner to the site that morning. He marked the find on his GPS. The ranger's wife emailed me two topo maps showing the discovery site. It was a half-mile past Zeledón, upstream of Negritos in a canyon. I'd walked its rim many times.

How could I have missed him?

CHAPTER 48

Sleeping in the Forest

Above El Doctor, May 21, 2016.

We would head to Costa Rica as soon as we could get on a flight. I left Virginia while Peggy waited for Jazz to send her passport by air express. Somehow, my flight escaped a volcanic ash-fall that shut down all other flights into San José. I made it to Puerto Jiménez the same day.

Ken and a Tico named Gerhardt met me at the Fiscal's office in Puerto Jiménez at five the next morning. Gerhardt is a lean and gentle multisport athlete who worked as a local fixer for *Missing Dial*. He could handle the jungle better than anyone who'd worked on the show and we had become good friends during filming.

Gerhardt translated, explaining that the discovery site was on a small creek called El Doctor. This was the uppermost tributary of the Negritos canyon that Steve and I had descended on rope. The creek was named for a doctor killed there many years ago in an airplane crash. Miners know it for its strange, focused winds that knock down trees remarkably often, like the one that nearly landed on the LTR crew above the Zeledón.

It was the day that *Missing Dial* would air its first episode. Aengus had returned to the Osa for the unexpected turn of events. The timing seemed suspicious to him. He hurried to Dos Brazos to capture what he could of the action. In Dos Brazos, production shoved cameras into people's lives, probing without asking, only checking that a release had been signed to excuse their intrusions. There's no poetry in reality TV, no doing more with less.

I was embarrassed to be part of it now, especially in Dos Brazos, where residents who'd seen the trailers were horrified by the portrayal of their village and had encouraged the miner to go into the jungle. The miner explored the one corner that no one had searched and nearest to where Jenkins met Roman. It was the dry season, when the uppermost little tributaries like El Doctor are traditionally accessed.

Ken, Gerhardt, and I arrived in Dos Brazos at dawn. Pancho, the patient ranger who had taken Thai and me into the Conte years before, led us to El Doctor. I wanted to meet the OIJ forensics team while it was still on site. The four of us raced along park trails through the forest to Zeledón. Struggling to keep up with my younger, healthier companions, I felt fat, old, hot, thirsty, and tired. The last two years had taken their toll on my health. But our mission was urgent and I pushed myself.

By eight we reached the campsite where Ole and Steve, Brad and Todd had all camped with me. Twenty minutes later, we were at the forensic team's camp on the ridge above El Doctor. The Fuerza prevented Ken and Gerhardt—employed by TIJAT—from visiting the site. I went down with Pancho.

Ken had heartily drunk Carson's Kool-Aid. "I'll only believe it wasn't foul play if there's money and his passport with his pack. Otherwise, someone put all this here—or killed him—either Joe or the Guichos."

Pancho and I met the team carrying Roman's remains and camping equipment up a steep muddy trail. The group was big, over a dozen people. Among them was the director of the OIJ and the two detectives who'd worked the case from the beginning. The OIJ dog handlers and Jorge from the embassy greeted me. Jorge introduced me to the OIJ forensic anthropologist, a polite young woman named Georgina who spoke excellent English.

Returning with the team to their camp on the ridge, I sat with Georgina. The rangers put down their packs. Some were loaded with clear plastic bags filled with recognizable items: the yellow sleeping pad and an orange lash strap I had given Roman. The strap was one of many we had at our house and used to strap gear to packs and packrafts. Seeing these things and a clear bag of his bones, I broke down in tears and I sobbed. Georgina comforted me as I turned away from the crowd. This was really him.

I composed myself. Georgina told me that they'd found many bones. She said that the equipment and bones had been washed downstream and trapped behind logs. The OIJ found his pelvis in his shorts, a femur under the log, his skull near his pack. A ranger found a live fer-de-lance in the creek bottom and speculated that snakebite had killed Roman. The green Salomon shoes that had looked so fresh in the photos were actually falling apart. One contained foot bones, Georgina said, but the other did not; he had at least one shoe on when he had died.

Importantly, none of the bones showed any trauma. Nobody had hit him over the head. Nobody had hacked him with a machete. There were no signs of bullets striking bone.

AS THE REST of the team ate lunch and prepared to head back to Dos Brazos, Pancho led me and Gerhardt down a steep muddy trail to El Doctor. Only inches deep and two feet wide, the creek twisted through black bedrock with tropical plants arching overhead. It was good walking and we made our way quickly to the discovery site downstream.

Pancho, who'd been one of the first rangers at the site the Friday before, stopped and pointed out where the stove and fuel canister had been found near a pile of grapefruit-sized rocks that looked like a good camp kitchen. A few yards farther, he said, was where Roman's machete had rested on the pea-sized gravel of the streambank.

I looked around. The canyon was somewhat wider here, where the gravel bed offered the only place wide and flat enough to camp. The canyon walls, while still steep, slippery, and eroding, pitched back to less than forty-five degrees. It was easy to see this as a place where Roman might decide to camp. There was water and a soft gravel bar to stretch out his yellow pad.

Downstream was a mass of logs and forest debris, sticks of all sizes, leaves and twigs caught up in a choke point between rocks and the broken crown of a forty-foot hardwood tree trunk. It had fallen, rolled into the creek, and left its buttress perched at the top of a small waterfall with its crown upstream.

Time, decay, and the action of floodwaters had left the crown a logjam of eight-inch-diameter limbs. Yellow crime scene tape boxed in the tangle of branches and sticks. Roman's bones, clothes, and camp were found partly buried in this debris washed downstream by six hundred rainstorms.

So, this is where my son died.

Taking in the chaos of greenery, I thought about our young atheist at the end of his biological being. Roman would have considered his absorption and acceptance by the rainforest a fitting disposition. He was part of the jungle now.

It was little wonder that I had been drawn to the area again and again. It was so near to where Roman had met Jenkins. I had walked the ridgeline above El Doctor—only two hundred yards away—more than a half-dozen times. I had smelled the foulness of death on the first trip with Vargas and Thai when I found a dead tamandua, just short of where the rangers now set their packs holding Roman's bones and gear. A quarter mile downstream of the logjam, below a series of waterfalls, was a drop that Steve and I had scrambled past after our exploration of Negritos's slot canyon. We had turned back there to explore Jenkins's tunnels rather than proceed farther upstream.

Young spindly trees grew in the red dirt above the fallen log in the creek bed. "This all looks like new vegetation," I told Gerhardt, waving my arms toward the saplings.

"The miner who found him thought maybe a tree fell down on his camp," Gerhardt said. "He thought that the blast from the falling tree blew the stove upstream."

After taking pictures of the scene, I signaled to Pancho and we returned up the hill. We all hiked as a big group to Dos Brazos, where Aengus waited with cameras rolling. Elmer, who owned the cantina at the end of the road and worried about tourism, came up to me and suggested that, because we had now found Roman, we should cancel the show.

I agreed with him. But it wasn't my show to cancel.

CHAPTER 49

Closure

Passport, money, machete, map: May 22, 2016.

We regrouped at the Fiscal's office in Puerto Jiménez. Roman's gear and clothing were laid out in a back room. All of it was muddy. Much of it looked rotten. Asking questions, I inspected it closely, piece by piece, handling it, weighing its meaning in my hands. The pack lid was found separated from the pack. Inside was an unopened bag of cookies labeled "Chiky Chips" and a package of Tang. "He hadn't starved," somebody pointed out.

His passport had been in the lid, too, and inside his passport were three colorful bills totaling $37 worth of colónes and the disintegrating folded remains of "the best map yet." Inside his pack OIJ had found his mosquito-net tent and some extra clothes; outside of it, they found his headlamp, his Visqueen tarp, and his sleeping pad. It appeared he had stopped and was perhaps making camp. His compass, too, had been on the ground, its bearing set at 240 degrees, the direction from there to the Rio Claro. It had probably been around his neck for reference.

The heavier, metal things lying in the stream bed had been upstream a few yards from the logjam: the machete, a green fuel canister about eight inches long and four inches across, and the unidentified green item in the photo. This last piece was now obviously part of the stove. Roman must have assembled the stove, which was then struck by a massive blunt object that snapped the burner from the canister and pinched and folded the threads on the burner's valve. The steel fuel canister was dented with a broad divot, as if something hard and large in diameter—perhaps a hardwood tree limb—had hit it with extreme force.

Like Ken, Jorge from the embassy had been sure it was foul play until he saw the site. Now it was clear: a natural death, an accident. Snakebite or tree fall, it was all speculation now. I preferred a scenario that matched all of the facts with as little anguish as possible.

IT SEEMED TO me that Roman had met Jenkins and his band of miners along the Zeledón on July 10 or 11, then hiked upstream after breakfast. He would have taken the left and better-used fork, following Jenkins's trail. He passed by the path to their tunnels and descended to the Negritos, having bypassed its canyon. From there he negotiated a series of waterfalls upstream for a half mile to where he died.

This is what closure feels like.

Peggy arrived the next day and we inspected Roman's things together. We spread them across a table outside the Pearl. Like I had, she handled and inspected everything, verifying Roman's presence in the material remains, seeing his passport's muddy and faded photo taken when he was a teen, his name and birthdate visible, his neat handwritten notes on a folded sheet of paper.

Lauren came into the Pearl. She called in her big voice across the room for everyone to hear: "This whole time, for two years, you always said that you knew your son—that he would never walk with a guy like Pata Lora. And you stood up to Dondee and Carson." She smiled her big smile. "And you were right all along. You did more than any parent could be asked to do, Roman. You went above and beyond." Speechless, I tried to thank her, but couldn't.

Aengus, who had made a six-part series based on a story that wasn't true, wasn't so sure. He took me aside. "Don't you think we should get the FBI in here to do a real investigation? Just to be sure?"

"No. I don't, Aengus. You haven't been to the site where Roman was found. If you had, then you would know that it's impossible that Pata Lora or the Guichos were involved. There's no conspiracy. There's no murder."

I looked to Ken for support. He shrugged. "Yeah, Aengus, I know, it's hard to believe, but the money and passport were there. Nothing was taken. . . ."

"But aren't you just a little suspicious that all this is happening now, when the show's coming out?" This was the producer who had hired the ex–DEA agent who convinced me my son had been murdered. The producer who then turned the murder into a titillating trailer shown on TV ten times a day, as I saw myself in a hotel room.

I erupted, emboldened by Lauren. "Look, Aengus, can't you

just let a grieving parent *be*? For two years I've felt like I've been held underwater. And for the first time I can come up for air and I can breathe. And you just want to shove me back down? *No, Aengus, I have had enough!*"

At that moment, I saw in Aengus what others had whispered. He had seemed to be on our side. Now I wasn't so sure.

The next morning, Peggy and I went to El Doctor. I held her hand as we slipped and slid down the steep muddy hillside, following the tracks left by the repeated passage of a dozen OIJ, Fuerza, and MINAE. Peggy ducked beneath the yellow crime tape and started digging, first with a small stick and then with a spoon, brought for our lunch.

The rangers wandered off. Hesitant at first, I slipped under the yellow tape and joined her, the now-familiar feeling of looking for sign of Roman washing over me as we searched for anything that might help us understand what had happened. From experience, I knew that seeing anything of his would bring him close to me again and touch my heart.

A sympathetic Costa Rican had given us each a long solid walking stick made of local wood. Peggy's was light but sturdy and an inch and a half across. Mine was heavy, longer, and thicker, made of a tropical hardwood called manu. We used our walking sticks as levers to move the eight- and ten-foot sections of log aside, digging underneath, pushing aside the sediment and debris, looking, but finding nothing. The OIJ had been thorough.

I pointed out the dead tree, the new growth. "It looks like a tree fell on his camp, doesn't it? Although some of the rangers think it was snakebite and found a terciopelo down here. What do you think, Peggy? Do you think somebody killed Roman here?"

"No. No way. Why would anybody be here in the first place?"

"Maybe somebody killed him and brought him here?" I prodded.

"Too much work. How would they get him down the steep hill? Cut him up and carry him? It's hard to walk here even without a pack. No, he died here. He was probably in camp or making camp and a tree fell, probably in the dark and he couldn't see to run out of the way. Lots of trees fall here. Like we saw near Dos Brazos. Or when you were here with Brad and Todd and the Learn to Return guys." She sounded as convinced as I felt that it had been a natural death.

Afterward we walked back to Dos Brazos to meet the miner who had found Roman. The miner said that the locals felt a kinship with Roman, because he had explored off-trail in a very challenging canyon and forest area, and he had done it without permission, against authority. The miner said that Roman had the spirit of the gold miners and they all admired him for that.

WE LEFT THE Osa for San José, where we joined a press conference with the OIJ and embassy. In front of a room full of media, I thanked the miners, the rangers, Cruz Roja, the OIJ, and the embassy, even all the people of Costa Rica for their big hearts and helpfulness. Afterward we met with Georgina, gave our blood samples for DNA testing, thanked everyone personally, and prepared to leave for home.

There was only one step left. In an office in OIJ's brooding granite building in downtown San Jose, we told a soft-spoken translator that Pata Lora's story wasn't true. He had never been with our son. We retracted the denuncia that the OIJ had prepared to arrest Pata Lora for murder.

TIJAT's producers had been right: the power of the camera is real. The effort to have Roman's case moved from missing persons to murder had been successful, thanks to Carson and *Missing*

Dial. But in the end, the media's search for sensationalism had left us all vulnerable to a schizophrenic's self-incrimination.

Almost two years after Roman's last emails had thrown Peggy and me into a valley of grief that darkened and deepened with time, we now found ourselves atop a small hill of relief rising up from the valley bottom. Roman had not been murdered. He had not waited for us to save him. He had probably died before any of us knew he was in trouble. Before I had even read his last words: "it should be difficult to get lost forever."

We had found him.

CHAPTER 50

Gather the Ashes

Clouds over Costa Rica, December 2016.

By August 2016, we heard from Georgina that Roman's dental records matched. In October, she sent the DNA results from the bone marrow sampled inside a tibia. The DNA showed conclusively that the tibia came from our son. Then the embassy wrote asking us what we wished to do with his remains. We agreed on cremation. At the end of November, we flew down to see the bones, collect the ashes, and pay the reward to the miner who had found him.

Tourism was down since Hurricane Otto had just hit Costa

Rica and the volcano had erupted again. We met Peggy's sister Maureen and her husband, Steve, at the San José airport. In the morning, Gerhardt picked us up and took us through thick traffic to the funeral home, where I paid for Roman's cremation. Then we flew to Puerto Jiménez, rented a car, and went to the bank, where Steve donated money into a Cruz Roja account. I had a big wad of $5,000 in American bills to give to the miner who had discovered Roman's remains. The $50,000 offer was only a ruse to get Pata Lora to talk. We had no intention to pay that sum.

The entire town served as a reminder of our two years of searching. Peggy and I pointed out the new Fiscal headquarters to Maureen and Steve. It had moved from Golfito to Puerto Jiménez about the time Jorge from the embassy took charge of the investigation, perhaps because of Roman's disappearance, but more likely because of crime's increase on the Osa.

We walked by the ballfield where we had studied young men's feet in search of Roman's Salomon shoes, the secondhand store where we had looked for Roman's gear and clothes, and the Corners Hostel where Roman had stayed. Doña Berta recognized us and came over to say that she was happy that we'd found Roman, clutching my hand in hers. At a restaurant where we had breakfast, Andres, who'd taken us to Cerro de Oro, said the same thing. Maureen even spotted Pata Lora at the grocery store. I made sure to avoid him.

We drove past the waterfront restaurant where I had called Peggy and told her, "Roman will probably be irritated I'm here," and we had both laughed but agreed it was the right thing to do, coming down. So many places triggered so many memories of being wrong so often about what had happened to our son.

Sitting there along the waterfront, a gringo nodded and smiled. He looked travel-worn, with curly beach-blond hair, a scruffy beard, and a flowered travel shirt. *Hmm, another local who recognizes us.* Nodding back, I realized it was our friend Chris Flowers from Anchorage. We had planned to meet up in Costa Rica, but not here and not now.

Chris had his boys with him: Cody, nine, and Cole, eleven. When his second son was born, Chris called to tell me the news. I asked his newborn's name and Chris said, "Cody," adding, "I just hope he doesn't change his name to 'Roman' when he gets older." We both laughed.

Chris and his boys followed us to Dos Brazos to pay the miner's reward. We bumped along the potholed road, past flooded fields and Brahma bulls. The recent rains from Hurricane Otto had ravaged the Rio Tigre's banks and eaten into the oil palm plantations. Even the road looked like it might fall into the river.

We went to Jenkins's place. He had a nice new house with a metal framed roof and white walls, a tiled floor, and two bedrooms that opened with doors rather than curtains. Jenkins's younger brother was there. Out of hospitality, Jenkins's wife, Gladys, and their teenage daughter passed out pink Nestlé's Quik to all.

Jenkins told us the weather had been bad for almost three weeks, leaving everybody out of money, like the jungle's birds and monkeys were out of fruit. He showed us the portable sluice box he had received as a tip from a client he guided. He looked a little tubbier than when we had last seen him. He said that he was like a little sausage in his shirt: "fat and happy" came to mind.

He had his new house, built by the government, and he'd made money in construction out at La Leona on the park boundary past Carate. Peggy and I had walked there six months earlier to tell him that Roman had been found and most likely killed at El Doctor by a fallen tree or a snake. We thought he should know, since he had been the last person to see Roman alive.

Jenkins said the town was happy. Most had seen the show

and everyone could see that Pata Lora was lying, and that I had conflict with Carson, who believed all of Pata Lora's story. Tourism was returning now that the rains had slowed. People in Dos Brazos had heard what I said on the news after Roman's discovery. It seemed to confirm what the consul general, Ravi, had relayed to me: that everyone had appreciated my gratitude toward Costa Ricans.

We talked about fathers and sons. Jenkins told how his father was part Nicaraguan Indian and could charm snakes with his touch and wrap them around his neck like a scarf. Jenkins said he read a lot about Christ but wasn't religious. I told him I liked Christ, too, and that I hoped there was a God.

Jenkins went with me to translate while I paid the miner who had found Roman, then took Jenkins home where Arnoldo was waiting. Arnoldo had hosted Pata Lora and Cody all those years ago. I greeted him, then hurried after Peggy and the others who had hiked into the hills on the Fila Matajambre trail. I caught them where they had stopped to watch a big millipede crawling across the forest floor.

Before it rained, Cole found a green and black poison dart frog and a yellow spot damsel fly that looked like a helicopter as it flew. Peggy spotted a tamandua, the small black-and-cream-colored anteater with a long prehensile tail. The handsome little animal had been walking along the trail when it reared up like a boxer just a foot away from her, brandishing its long razor-sharp claws before climbing up a slender tree to escape.

Somehow, through all the ordeal while looking for Roman, I had come to see the tamandua as a sort of spirit creature for him, ever since the first day when Thai and I had driven to Carate and seen one climbing along a fallen roadside cecropia tree.

We stopped at the supermarket for ice cream and tamales full of moist cornmeal and seasoned pork wrapped in a banana leaf, then headed for Carate. We made it only as far as the third river crossing. The water was high. We could have used the current to cross, but returning at the same level or higher would have been impossible.

We parked there by the water and watched a half-dozen dainty little squirrel monkeys with black-capped heads, white faces, and straight tails. They watched back, curious and gentle. Chris and I helped a young guy get his dirt bike across the deepest channel, where the wheels floated and the current threatened to pull it away.

Driving back to Puerto Jiménez, we inspected a freshly dead armadillo and watched through binoculars as white-scrotumed, cat-sized male howler monkeys roared at each other. Chris's boys delighted as we drove through a noisy flock of red-shouldered green parrots. It had been great to have the excitement and enthusiasm of Chris's boys, a throwback to our own tropical trips with our own kids, and I felt we had come full circle—almost. Almost, because what I would have really wanted was something like this with Roman's kids.

AFTER OUR TIME on the Osa, Peggy and I went back to San José to view Roman's remains before they were cremated and we could bring the ashes home. I had brought a poem that resonated with Roman and his life and his death and reading it aloud over him held some meaning for me and for Peggy, too, I hoped:

SLEEPING IN THE FOREST

I thought the earth
remembered me, she
took me back so tenderly, arranging
her dark skirts, her pockets

full of lichens and seeds. I slept
as never before, a stone
on the riverbed, nothing
between me and the white fire of the stars
but my thoughts, and they floated
light as moths among the branches
of the perfect trees. All night
I heard the small kingdoms breathing
around me, the insects, and the birds
who do their work in the darkness. All night
I rose and fell, as if in water, grappling
with a luminous doom. By morning
I had vanished at least a dozen times
into something better.

Its poignancy set me sobbing, the words capturing how I imagined his final time had been. The funeral director in his shiny black shoes and neat mustache slipped out the door, closing it behind him to leave Peggy and me alone with Roman.

We stood over a stainless-steel bin that held the physical remains of his life. The smell of decay touched our hands as we poked and probed at the bones, soiled and brown, held by the forest for nearly two years. The source of the odor, I could see, was the head of a tibia, cut clean by a saw for the marrow's DNA.

"It's a little smelly," Peggy said.

She was so brave, as always, holding her heart together more strongly than I, checking to be sure this was her little boy. Peggy looked closely at the teeth, holding her eyeglass case as if it were lips to see that the teeth were right.

"I don't have my phone. Do you have a picture?" She wanted to see his toothy smile.

I fumbled through my phone's photo stream but could find no

picture of Roman to use for an impromptu dental comparison. I thought of Peggy's strength when she bore him the day we had first seen him, how I came apart then, too. This was the same.

She grabbed the bones, pushing the skull this way and that to get a better view, to be sure it was him. I was already sure, had been sure, and said something like, *"What good would doubt bring here now?"* At first, she thought that these teeth, these bones, were not his. But the sealants on the teeth were there, she said, and when she finally found a chip that she recognized in one of the incisors she confirmed this was our son after all.

I was relieved that the funeral director had left so we could pore over Roman's bones, looking for something, a story, or maybe just that connection to him that I missed so achingly, that I still miss writing these words now.

EPILOGUE: MEAT, RAVENS, AND SEEDS

Back in Anchorage a few weeks later, we held a memorial for Roman on the winter solstice, December 21, 2016. He had often thrown summer solstice parties for his friends, with food, bonfires, wrestling, and storytelling. During the short daylight hours before people arrived, I shoveled and cleared snow, positioning four fifty-five-gallon burn barrels. Peggy handed Jazz and her cousins big earth-toned Christmas ornaments to hang from trees in the yard. I lit the barrels' firewood and friends arranged icy luminaries lit by candles that cast a golden glow across the snow. We had thawed twenty pounds of moose meat and two sockeye salmon to barbecue on our grill. Friends brought dishes that they arranged on tables overflowing with food in the living room. Our home was warm and crowded and felt full of love.

Using a friend's plasma torch, I had cut out images and symbols on the burn barrels. The cutouts of DNA helices, moose, insects, bicycles, and Dungeons & Dragons motifs glowed brightly in the darkness of Alaska's shortest day of the year. The barrels were hot and threw their welcomed heat at eighty of us gathered round under the cold, clear night sky of winter. Then, one friend or family member after another came forward to tell stories about Roman, stories from twenty-seven years of an adventurous, affectionate, and fulfilling life. Some stories made us cry. Most made us laugh. They all reminded us of the love we had for him and the love he had given us. It felt good to have brought him home, to share in the memory of his life with family and friends.

After the memorial, when everyone else had left and only family remained, I asked Jazz, "How do you think it went?"

She thought for a minute and said, "It was really good, Dad. Roman would have approved."

The year after Roman disappeared, Jazz moved in across the street from us. It was reassuring to have her there, to be family and close. While Jazz had always liked to remind us that she was the only normal one in the family, she had been raised with travel and nature, too, but after the Harding Icefield, she had called it quits with outdoor adventures that didn't include Peggy. The two have always been close and now with our adult daughter across the street, they shopped for each other at Costco, swapped fall jackets and winter coats, and talked and texted daily.

We ate dinner together frequently, me barbecuing moose to Jazz's liking on the grill that she'd bought me for Father's Day. "I want some of that finger meat," she'd say, requesting the gristle and tendon that came thick with red meat I trimmed from the "guest cuts." Salted with rub and chewy, finger meat requires that you hold it by hand while pulling the meat free with your teeth, like a piping hot piece of jerky but juicy. Jazz had helped cut and wrap the moose when I returned home from the hunt.

While Peggy and I had been away in Costa Rica for weeks at a time, Jazz took care of our house and yard, watering the grass and Peggy's greenhouse plants, checking in after a big earthquake had rocked south-central Alaska. She is reliable and capable, and we are proud of her. She'd been promoted from office manager to comptroller, earning a raise at the place that she'd worked for five years. The couple who owned the business liked her so much that they paid for her MBA, too. I had taught both our kids a year of calculus at APU when they were in high school. Helping Jazz with statistics for her business classes satisfied me and she seemed to enjoy the time together, too, typing code into her computer on our kitchen table.

Our family felt so much smaller, like a body missing a limb, but it felt just as close as it had ever been. Maybe closer.

IN THE YEARS following Roman's disappearance, I made a number of short day trips around Anchorage, mostly whitewater paddling. Home from the Darién Gap, I packrafted the Grand Canyon with Brad, Ganey, Steve, and another close friend who'd been in Veracruz. The trip had been a welcome diversion from my heartbreaking inability to find my son. Six years earlier, Roman, Gordy, and I had also packrafted the Grand Canyon, an experience that was heavy in my heart during this trip, just six months after his last email. Paddling the big rapids or joking with my companions was a healthy distraction. But when our boats drifted apart, leaving me alone below the red rock walls, deep in the inner gorge's calmer pools, melancholy crept in as I confronted the time Roman and I had spent there.

Since first moving to Fairbanks as a teenager, I have lost many friends to adventure. But the pain of losing Roman has gone far, far deeper, deeper than any pain I have ever felt, physical or otherwise. During the two years of searching, I couldn't bear to be alone with my thoughts, which always circled back to what might have happened to Roman and what I could have done somewhere, sometime, somehow in his life to have prevented his disappearance.

Even little things in my past could somehow warp into a cause. The questions I asked myself in Costa Rica—*Was I responsible? Would I have raised him differently? Had I paid close enough attention? Had I been too selfish?*—are questions I still wrestle with, and perhaps always will. But I know that, like the four most famous lines in Tennyson's poem "In Memoriam, A.H.H.," the bond we had was better to have than not. Would I have raised Roman the same way knowing that he would die on a path I led him along? The answer is obvious but the question unfair. We never know the future. There was no single moment in Roman's upbringing that can be traced forward to his death, no chain of events, no cause and effect. Accidents happen. Time has passed, and while these questions no longer crowd my heart, they linger.

Eventually I did manage a solo trip, my first since his disappearance. I have never been much of a soloist, although I have made trips alone over the years. In September of 2017, I went to Nuuk, Greenland, its capital, for a scientific conference at which Ganey and I would both give talks related to his thesis research. Arriving early, I took my packraft and went out for a few days by myself.

I paddled with the tides through Greenland's coastal fjords, a magical, stark landscape I'd never seen before. Flocks of eider ducks dove under the water in unison when my boat was too close. Overhead a peregrine falcon chased a huge deep-winged white-tailed eagle, bigger than any Alaskan raptor. A raucous group of eight young ravens followed me in my boat for an hour. One carried a sea urchin in its bill. The bird dropped the echinoderm to break and eat it, solving the mystery of how all the sea urchin shells had made it so far up on the hillsides I had walked across for two days.

Greenland in September felt a bit emptier of life than Alaska's arctic, but the going was easy enough and I had ample time to reflect, surrounded by barren tundra and the fjords' calm seas. Of course, I thought often of Roman. He had loved sea-kayaking the bays of Alaska's Prince William Sound. Greenland's fjords and arctic tundra would have piqued his interest instantly. He would have offered a sharp analysis, brimming with comparisons to the Alaskan waterways he knew and loved, both as an explorer and as a scientist.

It was easier now than in the previous three years, to be somewhere new without him, but it was still painful, like a bruise that doesn't heal, perpetually tender to touch. There was so much I wanted to tell him, things about the indigenous Greenlandic people of Nuuk, or the ravens that followed me, or the thousand other little details I drifted past.

I wished he'd been just an email away to share the new facts I'd learned at the conference on polar and arctic microbes. He would have found the red-colored bacteria that nucleate hailstones fascinating, the organisms that live in the salty brine of the ice pack improbable. A critical piece of Ganey's experiment had been Roman's suggestion. And he had been with the Japanese, Jazz, and me on the Harding Icefield looking for red snow, half his short lifetime before.

Roman and I were so close. Paddling alone in a wilderness he would have loved, I discovered that I was slowly learning to live with this chronic injury set deep in my soul. As I paddled farther into the Arctic, thoughts of him invaded every crevice of my life. They still do, where they germinate and grow like dropped seeds.

ACKNOWLEDGMENTS

A handful of writers were invaluable in organizing my writing, particularly Michael Wejchert in several chapters of Part I. His skill in selecting nuggets from my stream-of-consciousness recollections and ferreting out confusion allowed me to rewrite my old stories coherently, accurately, and truthfully. He was an excellent coach and structural editor for Part II. Our discussions in my office encouraged my way forward in Part III, most of which I'd written while in Costa Rica, Panama, Alaska, and D.C. as events unfolded. I am grateful to the young writer for his advice. He was the book's first reader and particularly adept at helping me see what was missing.

From the beginning, David Roberts has been the driving force behind my book. For decades he has encouraged me to write a memoir, but it wasn't until my son went missing that I needed to tell a story at book length. By sheer happenstance we encountered one another in a remote Arizona canyon along the Mexican border in March 2017. At first, I thought Roberts—dressed in khaki shorts and a floppy sun hat—was just another springtime birdwatcher. Drawing closer, I saw through the willows it was he and Sharon balancing on boulders, picking their way upstream on a research expedition for yet another of his wonderful books.

Roberts invited Peggy and me to dinner in Tucson and later to the Airbnb that he and Sharon were renting. Over a bottle of merlot, he convinced me I needed to write this book. Given the auspicious serendipity of our encounter midstream along a desert creek bottom, Peggy insisted I move forward. I soon set to writing in earnest. Roberts read early drafts, reviewed grammar, and like a psychoanalyst pushed me to explore uncomfortable memories where important themes dwell.

I'd also like to thank two Alaska Pacific University faculty for each reading an early draft: David Onofrychuk and Mei Mei Evans. Like Michael, David was particularly adept at encouraging me to write more clearly and succinctly, pointing out redundancies and narrative side trips. As parents and writers, David and Mei Mei also forced me to confront the heart of this book. I am fortunate to have colleagues like them who take time to help me.

Gordy Vernon is another writer I admire and respect. Michael usually asked for more words, more sentences, more paragraphs; Gordy less. Jon Krakauer offered support and structural suggestions. I appreciate his advice. Michael, David, Gordy, and Mei Mei all made suggestions that chaffed me at first and I let them know. But in the end, I followed (almost always) their advice. Maybe I owe them my apologies as well as my thanks.

Peter Hubbard, my editor at William Morrow, has been a true pleasure since the moment I met him. His fifteen years at HarperCollins show and he has been insightful, helpful, and gentle. It would be wonderful to work with him again. Nick Amphlett took care of the many necessary details to make this a book. I appreciate the work they put into publication, going so far as to have the copyedited manuscript airdropped to me in the Brooks Range wilderness.

My agent, Stuart Krichevsky, went far beyond what I would expect any agent to do. With his team, including Laura Usselman and Aemilia Phillips, he helped me arrive at a place where my notebooks, stories, and ideas became this book. I am especially grateful to Stuart for having faith in me, given the early words I provided.

Without Peggy's support this book would simply have been impossible. Not only did she encourage me to write, but she sacrificed her time to support me while I did. She literally took care of me during the months I did nothing but write from when I woke to when I slept. Our daughter, Jazz Dial, too, supported my efforts. Without these two I would be lost.

Friends and family who read early drafts and caught typos, provided comments, and recalibrated my recollections include: Peggy, Jazz, Steve and Maureen Haagenson, Linda Griffith, Tamara Dial, Lauren Cleaver, Thai Verzone, Brad Meiklejohn, Dick Griffith, Carl Tobin, Chris Flowers, Jon Underwood, Nancy Brady, Paul Twardock, and Michael Martin.

This book is dedicated to the family, friends, friends of friends, former and current APU students, U.S. and Costa Rican officials, acquaintances, miners, rangers, Cruz Roja volunteers, OIJ, U.S. Embassy, FBI, TIJAT and its contractors, and even strangers who came forward to help us and support us physically, financially, emotionally, and spiritually. All of them helped Peggy and me find Roman.